ZOOLOGY MADE SIMPLE

ZOOLOGY
MADE SIMPLE

DOROTHY F. SOULE, M.A.

Research Associate in Zoology

Allan Hancock Foundation University of Southern California

MADE SIMPLE BOOKS
DOUBLEDAY & COMPANY, INC.
GARDEN CITY, NEW YORK

ABOUT THIS BOOK

It has been said that more than half of what comprises the study of zoology today was unknown ten years ago. While one might disagree with the precise proportion and the time interval, it cannot be denied that a revolution—or evolution—has occurred in the field of zoology in recent years. The subjects of biochemistry and biophysics, for example, have had great impact on zoological knowledge, and the electron microscope has revealed details of ultrastructure heretofore totally unknown. Teachers and authors are constantly faced with a multitude of new information that must be integrated with the established knowledge. They must re-examine old concepts and make difficult choices of what to include or omit from lectures and texts that impose rigid time and space restrictions.

I hope this book will serve to impart some of the wealth of information available on a general level. Perhaps it will also encourage a few to delve more deeply into the phases that particularly interest them. Short of these goals, perhaps some will at least gain an appreciation of the scope of zoology and the nature of the efforts of those who are firmly committed to the field.

My deep appreciation goes to my husband and two daughters, whose interest, encouragement, patience (and sometimes typing abilities) have enabled me to complete this manuscript. Various colleagues, including my husband, have been kind enough to read critically appropriate sections; for this invaluable aid I am thankful, although they must not be held responsible for any errors of commission or omission. My thanks especially to Drs. P. H. Wells, John S. Stephens, William Hovenitz, and Dudley Thomas for manuscript suggestions.

—Dorothy F. Soule

CONTENTS

ABOUT THIS BOOK 5

CHAPTER 1

INTRODUCTION 13

History 13
Nomenclature 13
Golden Period of Ancient Greece and
Egypt 15
Rome 16
The Scientific Renaissance . . . 16
Development of Naturalism . . . 17
Rise of Rationalism 17
Conclusion 17

CHAPTER 2

THE ORIGIN OF LIFE 19
Spontaneous Generation . . . 19
Conflicting Theories of Generation . . 20
Formation of the Earth . . . 20
Development of Life 21
Atoms, Molecules, and Elements . 21
Atoms 21
Forming Simple Molecules—
Inorganic Synthesis . . . 22
Properties of Simple
Molecules 23
Organic Compounds 24
Inorganic Compounds . . . 26
Origin of Organic Compounds . 26
Carbohydrates 26
Fats 26
The Role of Nitrogen . . . 26
Origin of Nitrogen Compounds . 26
Importance of Ammonia . . 27
Peptide Linkages 27
Formation of Proteins . . . 28
Proteins as Enzymes . . . 28

Purines and Pyrimidines . . . 28
Phosphorus and Phosphates . . 29
Nucleosides and Nucleotides . . 29
Nucleic Acids 29
Ability to Replicate or
Reproduce 30
Primordial Energy 30
Physical Properties of Matter . . 31
Colloids 32
Nucleoproteins 32
The Changing Atmosphere . . 33
Chemosynthesizers 34
Bacteria and the Nitrogen
Supply 34
Autotrophs 34
Photosynthesis 34
Heterotrophs 35
A New Atmosphere with Oxygen . . 35
Transitions 35

CHAPTER 3

SIMPLE LIVING ORGANISMS . . . 37
The Unicellular World . . . 37
Characteristics of Living
Organisms 37
Viruses: Living or Nonliving? . . 37
The Monera 37
The Protistans 38
Volvox 39
Euglena 39

CHAPTER 4

THE CELL 41
Development of the Cell Theory . . 41
Early Cell Observations . . . 42
Ultrastructure of Cells . . . 43
Electron Microscope . . . 43

The Shape and Size of Cells . . . 43
The Cell Membrane 43
Food for Cells 43
The Nuclear Membrane . . . 45
Cytoplasmic Structures . . . 45
Functions of Cells 46
Unicellular Functions . . . 46
Multicellular Functions . . . 47
Cell Linkage in the Sponge . . . 48
Specialization of Tissue 48
Organs and Organ Systems . . . 48
Histological Specialization 49
Functions of Cells 49
Background of Histology . . . 49
Techniques of Histological
   Study 49
Structure of Tissues 50
Basic Tissues 50
Epithelial Tissues 50
Simple Epithelium Tissue . . . 50
Stratified Epithelium Tissue . . . 51
Glandular Epithelium 52
Connective or Supportive Tissues . . 54
Fibrous Connective Tissues . . . 54
Basic Connective Tissue Cells . . 54
Other Connective Tissue Cells . . 54
Fibers 55
Ground Substance 55
Loose Connective Tissues . . . 55
Dense Connective Tissues . . . 56
Cartilage 57
Osseous Tissue 58
Blood 58
Muscle 59
Cells of the Nervous System . . . 60
Conclusion 62

CHAPTER 5

REPRODUCTION AT THE CELLULAR
   LEVEL 63
Mitosis—Life from Life . . . 64
Phases of Mitosis 64
Prophase 64
Metaphase 65
Anaphase 65
Telophase 65
Interphase 65
Mitosis in Plants 66

The Nature of Genes and
   Chromosomes 66
The Genetic Code 68
Structure of DNA 69
Replication 69
Amino Acid Codes 70
Differentiation 71
Bisexuality—A Successful
   Adaptation 73
Meiosis 75
Preleptene 76
Leptotene 76
Zygotene 76
Pachytene 76
Diplotene 76
Diakinesis 77

CHAPTER 6

HEREDITY 79
Mendel's Contribution . . . 79
The Concept of Dominance . . . 79
Heredity Theories 82
Variation 83
Modern Genetics 84
Sex Determination 84
Nondisjunction 85
Back Crosses 87
Polygenic Inheritance 88
Mutation 89
Enzyme Action 89
Blood Types 90
Population Genetics 92
Eugenics 95

CHAPTER 7

ORGANIC MAINTENANCE 97
Elements Required by Living
   Systems 97
Nitrogen Fixation 98
Diffusion 99
Osmosis 100
Metabolism 102
Amino Acids 102
Vitamins 102
Fat-soluble Vitamins 102
Water-soluble Vitamins . . . 103
Digestion 105

Stomach 105
Small Intestine 105
Pancreas 106
Large Intestine 106
Intracellular Digestion . . . 106
Bond Energy 106
The Oxidation-Reduction
Mechanism 107
Metabolism of Glucose . . . 108
The Krebs Cycle 110
Reproduction 111
Methods of Reproduction . . . 111
Survival 111
Protective Coverings . . . 112
Copulatory Organs . . . 112
Nutrition 112
Larvae 112
Egg-laying 112
Germ Cells 113
Gonads and the Primitive
Kidney 113
Mammalian Reproduction . . . 114
The Female Organs . . . 114
The Male Organs . . . 115
Embryonic Sexlessness . . . 115
Descent of Testes . . . 115
The Estrus Cycle . . . 116
Female Hormones . . . 116
Fertilization 117
Cleavage 118
Gastrulation 119
Amphioxus Gastrulation . . . 120
Amphibian Gastrulation . . . 122
Gastrulation in Fish and Birds . . 122
The Primitive Streak . . . 123
Mammalian Cleavage and
Gastrulation 123
The Notochord 124
Protective Membranes . . . 124
Chick Amnion Formation . . 125
Mammalian Membrane
Formation 125
Determination 126
Differentiation 126
Epithelial Tissue Behavior . . 126
Mesenchymal Tissue Behavior . . 128
Vertebrate Morphogenesis . . . 128
Defective Development . . . 129
Developmental Anatomy . . . 129

The Nervous System 129
Nerve Transmissions . . . 130
Nerve Cell Differentiation . . 131
Neural Crest Cells . . . 131
Medulla 133
Cerebellum 133
Midbrain 133
Diencephalon 133
Optic Chiasma 134
Telencephalon 134
Cranial Nerves 134
Olfactory Nerves . . . 135
Optic Stalk 135
Vagus Nerve 136
Autonomic System . . . 136
The Digestive System 136
Visceral Arch Derivatives . . 137
Pouch Derivatives . . . 137
Thyroid Gland 137
Tonsils 137
Carotid Bodies 138
Thymus 138
Parathyroid Glands . . . 138
Postbranchial Bodies . . . 138
Lungs 138
Esophagus 139
Stomach 139
Intestine 139
Digestive Glands 139
The Circulatory System . . . 139
Aortic Arches 139
The Venous System . . . 140
The Heart 141
Folding 141
Lung Circulation . . . 141
Warm-bloodedness . . . 141
Embryonic Circulation . . . 142
Changes at Birth . . . 142
The Lymphatic System . . . 143
Somites 143

CHAPTER 8

THE ANIMAL KINGDOM 145
Symmetry 145
Organs Present 145
Tissue Complexity 145
Body Cavity 146
Coelom Formation 146

Intermediate Groups	146
Other Characteristics	146
Phylum Protozoa	147
History	147
Morphology	147
Physiology	150
Reproduction	150
Organelles	150
Ecology	151
Parasites	151
Phylum Mesozoa	153
History	153
Physiology	153
Phylum Porifera	154
History	154
Morphology	154
Classification	154
Physiology	154
Embryology	154
Ecology	154
Phylum Coelenterata (Cnidaria)	155
History	155
Morphology	155
Physiology	155
Embryology	156
Phylum Ctenophora	156
Physiology	156
Embryology	157
Ecology	157
Phylum Platyhelminthes	157
Origins of the Bilateralia	157
Morphology	157
Cestoda	158
Trematoda	158
Physiology	158
Embryology	159
Life Cycles	159
Phylum Nemertina	159
Phylum Aschelminthes	160
Rotifera	160
Ecology	161
Gastrotricha	161
Kinorhyncha	161
Nematoda	161
Nematomorpha	162
Phylum Acanthocephala	162
Phylum Entoprocta	163
Phylum Annelida	163
Hirudinea	164
Annelid Allies	165
Phylum Echiuroidea (or Echiurida)	165
Phylum Sipunculoidea	165
Phylum Priapulida	165
Phylum Mollusca	166
Amphineura	167
Gastropoda	167
Pelecypoda	167
Scaphopoda	168
Cephalopoda	168
The Lophophorate Phyla	169
Ectoprocta	169
Phoronida	169
Brachiopoda	169
Phylum Arthropoda	169
Class Crustacea	170
Class Insecta	171
Myriapod Arthropods	171
Subphylum Chelicerata	173
Phylum Chaetognatha	173
Phylum Echinodermata	173
Phylum Hemichordata	174
Pogonophora	174
Phylum Chordata	174
Conclusion	176
Ecology	176
GLOSSARY	179
INDEX	187

ZOOLOGY MADE SIMPLE

CHAPTER 1

INTRODUCTION

Zoology is, simply, **the study of animal life.** In a sense, everyone is a bit of a zoologist, for we have all observed animals, including our fellow men, throughout our lives, and have thus formed conclusions about them. We see that animals come in a wide variety of sizes, shapes, and forms. We note animal activities, such as eating and sleeping, and observe that animals react to their surroundings by showing pain, anger, and fright much as we do ourselves. We observe that animals reproduce themselves, each in his own form and no other—like begets like. We are usually unable to go beyond casual observations, however, because to do so it is necessary to master purposeful experimental techniques and make controlled observations in order to ascertain less apparent information about animals. In short, to study animals, we use an objective method—the science of zoology.

HISTORY

The science of zoology has undergone many changes since the early Greek philosophers prepared some of the first written scientific discussions in which they made some surprisingly accurate observations regarding animal description. But they spent far more time speculating and evolving highly imaginative theories than they did in dispassionate observation, with the result that much that they recorded was far from factual. Not until well after the Middle Ages and the Renaissance in Europe did people begin to make accurate observations and perform controlled experiments, and to question the dogmatic "truths" that had been established by the ancients.

The seventeenth and eighteenth centuries saw a period of increased freedom for the expression of scientific curiosity and the voicing of new ideas. The scientific method was introduced and applied, experiments under controlled conditions were made, and, more important, questions were asked with no prejudice as to the answer which might be reached. Even today, one of the ever-present menaces to good experimental technique is the orientation of trying to make the experimental evidence fit a pre-conceived "answer" instead of accepting whatever answers the tests may produce.

The formal study of zoology has passed through many phases. The first was essentially one of enumerating all of the various animals that could be discovered, describing their forms, and naming them. The **morphology,** or the description of form, of animals is a basic feature of traditional zoological study. A natural adjunct of this description is the naming of the animals described, for there are well over a million different kinds, or species, of animals. Without some method of naming and grouping animals it would have been impossible to co-ordinate the information that was gathered. Obviously, some animals resemble others rather closely, while others are completely different in appearance, so that grouping similar animals is a logical step. Aristotle, in the fourth century B.C., formulated the first method for sorting animals. Individual forms, such as the horse, cow, dog, or lion, were called *eidos;* all combinations of a higher degree were called *genos.* Aristotle formulated the science known as **taxonomy,** or classification. He said: ". . . animals may be characterized according to their way of living, their actions, their habits, and their bodily parts."

Nomenclature. Aristotle's system was followed for centuries until the Swedish naturalist Carl von Linné (known as Carolus Linnaeus because all learned works were then written in Latin) published his *Systema Naturae* in 1735, bringing long-overdue organization to taxonomy. Linnaeus standardized the **binomial system of nomenclature** in which each animal receives two names, the first name being that of its **genus** or group, and the second, the name of its **species.** Thus, the tiger is *Felis tigris,* and the domestic cat, a relative, is *Felis catus.*

The species concept is the basis of the classification of animals. A group of animals which are identical within the limits of recognizable variability, and which interbreed freely, are considered to constitute a species. A genus (*pl.:* genera) is usually a group of several species which resemble each other fairly closely and which may occasionally interbreed. This latter distinction is not always easy to determine, especially in rare animals or in animals with habitats that make it difficult to observe them, such as the environments of marine organisms. Genera are, in turn, grouped into **families** on the basis of some similarity; families are grouped into **orders;** orders, into **classes;** and classes, into **phyla.** In-between stages using the prefixes sub- and super- with any of these levels may be set up.

In taxonomy, the various divisions into which animals or plants are sorted are called **taxa** (*sing.:* taxon). Arranging organisms into a logical set of groupings that reflect their relationships to one another and to their ancestry is also known as **systematics.** This term is based on the title of Linnaeus' work. Some authorities attempt to differentiate between taxonomy and systematics, but the words are used interchangeably by many.

Since Linnaeus' time a number of international congresses of biologists have been held where the formulation of uniform rules of nomenclature has been discussed. A code has been developed so that everyone in the field will apply the same principles in naming and describing plants and animals.

Botanists have agreed on endings that will be applied to all the names used in a given taxon. Zoologists have not yet agreed on endings above the level of families, but many follow suggested forms. It would be quite a task to change many of the old accepted names to conform to new rules. A comparison of the suggested zoological endings and those in use by botanists, together with a listing of taxa, is shown in Table I. It will be seen that the required groups are present: phylum, class, order, family, genus, and species. Intermediates are also given, but it is not required that all be used. Some authorities have developed other taxa but they are not commonly used.

TABLE I

Linnaeus' Taxonomy

ZOOLOGY		BOTANY	
RANK	ENDING	RANK	ENDING
Kingdom Animalia		**Kingdom Plantae**	
Subkingdom		Subkingdom	
Superphylum			
Phylum		**Division (Phylum)**	*-phyta*
Subphylum		Subdivision	*-phytina*
Superclass			
Class		**Class**	*-phyceae*
Subclass		Subclass	*-phycidae*
Superorder			
Order		**Order**	*-ales*
Suborder		Suborder	*-ineae*
Superfamily	*-oidea*		
Family	*-idae*	**Family**	*-aceae*
Subfamily	*-inae*	Subfamily	*-oideae*
Tribe	*-ini*	Tribe	*-eae*
Subtribe	*-ina*	Subtribe	*-inae*
Supergenus			
Genus		**Genus**	
Subgenus		Subgenus	
Superspecies		Section	
		Subsection	
		Series	
		Subseries	
Species		**Species**	
Subspecies		Subspecies	
Variety		Variety	
Race		Subvariety	
		Form	
		Subform	

Phyla (*sing.*: phylum) consist of closely allied animals which share similar characteristics of morphology, function, or mode of development that are not seen in any other group, at least not in the same combinations.

At the time Linnaeus began his studies, the doctrine was widely held that every single kind of animal and plant had been individually created by a supreme being. Linnaeus had been taught this "doctrine of special creation." Still he could not avoid seeing the close similarities of some of the animals, so he indicated that perhaps the generic type was the basic one formed and the variations of the species had developed from them. Linnaeus predated Darwin and the theory of evolution by a century; yet he indicated that some sort of change was too obvious to ignore. Sweden was a conservative country in his time and Linnaeus was often in trouble for his unorthodox ideas. It is said that it took the intervention of the king himself to get him out of trouble when he dared to state that he believed there were two sexes in plants as well as in animals.

The system begun by Linnaeus has been of tremendous help to an orderly development of knowledge; this does not imply that some other system could not be used. Currently, there are zoologists who advocate a numerical system and the application of computer techniques to the sorting of organisms. Perhaps eventually every animal will have its own "social security" number. It must be kept in mind, however, that the sorting devices are artificial, set up arbitrarily by man, and represent attempts to classify that which actually exists in nature. In the early twentieth century, it was the vogue to state flatly that there existed nine, or twelve, phyla and to sort all animals into these pigeonholes. It is now recognized that this is not possible; various authorities of equal eminence may disagree on the exact relationships of certain animals, and their positions must sometimes be altered in the light of more recent investigations.

Studies in morphology are concerned with both internal and external anatomy. Some of these studies date back to medical writings on papyrus of the ancient Egyptians a thousand years before Aristotle.

Anatomical studies can be traced to early primitive society in which an unidentified citizen, using hot coals as a medium on the cave walls, sketched remarkably accurate representations of bison and elephant, carefully indicating the vulnerable areas on the animal which when hit with the arrow would bring death. The Minoan prehistoric culture of Crete, centuries before the golden age of Greece, showed signs of accurate physical observations. The Greeks, no doubt, drew some of their anatomical knowledge from the old civilizations of Mesopotamia and Egypt. One of the earliest known medical studies, recorded on papyrus in Egypt, tells of treatment for a "fallen womb." Symbols which later became hieroglyphics were diagrammatic illustrations of the uterus, the heart, the trachea, and the lungs, and dated from as early as 2900 B.C.

The Greek Alcmaeon (about 500 B.C.) began the dissection of animals, working toward the creation of an organized medical science. He even attempted some embryological observations. In Sicily, Empedocles (495–435 B.C.) influenced many students who came later with his statement that the blood is the seat of life. Many of the early Greeks believed that the heart was the center of thought and feeling, and that the arteries carried "air" (or *pneumos*, a vital force) as well as blood. Empedocles is sometimes considered to be the first proponent of evolution since he tried by hypothesis to explain the gradual development of different kinds of animals, although he based his theories on a kind of spontaneous assembling of random parts.

Actually, the Greeks were quite vitalistic in their philosophy about man's close relationship to his environment; they did not exhibit the ideas of any unique character for man as later cultures did. The Greek philosophers and metaphysicians devoted many years in search of the laws of nature. Because they were so limited in their opportunities for observation, experimentation, and exchange of information, their writings were often speculative deductions.

A group of medical writers on Cos in western Asia Minor during the fifth century B.C. became associated with Hippocrates, "the father of medicine," whose name is still associated with the physician's oath. Collected writings of the group cannot be attributed to any one author and Hippocrates is himself a shadowy figure, but the ideas the group set forth markedly influenced science for two thousand years. Here is found the idea of the four elements which compose nonliving matter: earth, air, fire, and water. The "humors" which correspond to these in living matter were: black bile (melancholy), yellow bile (choler), blood, and phlegm (pituita).

GOLDEN PERIOD OF ANCIENT GREECE AND EGYPT

Athens reigned as a center of learning from about 400 B.C. until 290 B.C. Socrates, the great philosopher, and his pupil Plato did not favor investigative science and Plato published some entirely fanciful statements on the human body. Because of his eminence, however, his work, good and bad, exerted a strong influence during the Middle Ages. Aristotle (384–322 B.C.), brilliant light of the ancient world, compiled the scientific knowledge of his day into three works: *History of Animals, Generation of Animals,* and *Parts of Animals.* He developed the science of comparative anatomy, describing the anatomy of many animals. He studied the embryological development of various animals and developed a scheme of classifying animals on this basis. It is tempting to digress at length into these records of antiquity; to see, for example, how science died at Athens with the loss of liberty and the fall of the Alexandrian Empire.

In Egypt the Ptolemaic kings encouraged learning and established a library and a museum. Two Asiatics, Herophilus and Erasistratus, became the greatest anatomical teachers in Alexandria. Herophilus (about 300 B.C.) dissected and compared both human and animal bodies, regarded the brain as the site of intelligence, and distinguished between veins and arteries and motor and sensory nerves. Erasistratus (290 B.C.) is credited as the founder of the study of physiology, as Herophilus is credited with formalizing the study of anatomy.

Erasistratus came very close to discovering the circulation of blood, and described the smallest vessels which connected veins and arteries (the capillaries). He missed the actual nature of the heart pump because he believed the lungs brought to the heart a "vital spirit" which was sent then to the brain by the arteries. There it became "animal spirit" which was distributed by the nerves.

ROME

The absorption of Egypt into the Roman Empire, and the end of the Ptolemaic dynasty at the death of Cleopatra in 30 B.C., brought also the end of Alexandria as a center of scientific thought.

The Roman Empire produced no climate of free inquiry, and consequently little of scientific worth was accomplished. Pliny (A.D. 23–79) was widely read during the Middle Ages; yet his voluminous ramblings were of little or no value. Galen (A.D. 130–200), foremost anatomist of his day, based his writings primarily on the Hippocratic collection and Aristotle's work. However, he became preoccupied with religious-philosophical bias, and his work discouraged further investigation for centuries. He expounded a "perfect creation" of every part by a god, not clearly either the deity of the pagans nor yet of the Christians, and advanced a "divine purpose" for each organ. The ascribing of human purposes to

structures is known as **teleology,** and still arises to plague modern scientists.

THE SCIENTIFIC RENAISSANCE

The nonproductivity of the Middle Ages is well known, its causes less well understood. There is no single reason for the lack of advancement; a number of factors working together affected learning. The great plagues that decimated the population several times made living on the barest minimal standard a struggle and discouraged communication because of the fear of contagion. The church has sometimes been blamed for restricting the progress of science by its emphasis on tradition and authority. Yet it was the church and its monasteries that preserved the libraries and recopied the manuscripts of antiquity, saving them from complete destruction. It was not until the sixteenth century that independence of thought and observation began to stir and science was revived.

Andreas Vesalius, born in 1514 in Brussels, was to become the master anatomist of history, for his was the first outstanding contribution to modern science based on original observation. The Renaissance changed the intellectual climate, and Vesalius was certainly a product of this age. He was only twenty-nine when *Fabrica,* his seven-volume work on the human body, was published. He had studied in France and had already become an important professor at the great University of Padua, but although he lived until 1564, he never produced further research of any consequence. In 1543, the same year in which Vesalius' *Fabrica* was published, Nicholas Copernicus, the Polish astronomer, published his work on "the revolutions of the celestial spheres" which refuted the ancient concept of the earth as the center of the universe by describing the sun as the center of the planetary system.

Some of the scientific works of the period did not fare as well as those of Vesalius. In Paris, a Spanish fellow student, Michael Servetus, discovered the pulmonary circulation in man and approached the idea of blood circulation. However, he incorporated this and other ideas in a religious book which he had printed secretly. He was tried and condemned and burned at the stake, along with all but three of the books, by the medieval Inquisition in 1553. Bartolomeo Eustachi, who practiced medicine near Rome, knew Vesalius' works and improved on some of them, making some excellent drawings. His work was published posthumously: it was not until 1714 that the plates of his remarkable set of anatomical drawings were found and printed although the written works remained lost. The impact by then was much less than if the works had been published contemporaneously with those of Vesalius.

DEVELOPMENT OF NATURALISM

The philosophical naturalists began to flourish during the sixteenth century, almost as a secondary result of the change in art from the strictly religious to the naturalistic portrayal seen in the Renaissance. This was in turn probably connected with the increase in travel that had begun in the fourteenth century. The culture of the Greeks, which had disappeared from the Roman Empire, had been kept alive in the Near East, translated into Arabic in many cases, throughout the Middle Ages. Travelers began to return with the writings of the classical scholars; this caused a flurry of translating to scholarly Latin. Ironically, there was also great fervor in deleting any contributions the Arabian scholars had made. The careful naturalistic illustrations and sculpture of the ancient age were much admired and copied by both artist and scholar. In order to make the works more widely available, they were often copied, until finally some of the copied illustrations bore no resemblance to the originals. Only then did the artist and naturalist fortunately realize that they, too, might follow the classical example and observe nature accurately for themselves. For years there lingered this devotion, for example, to Aristotle; it was a common pastime among the educated to try to identify all the plants seen in northern Europe according to Aristotle's descriptions of his native flora in Greece—a ridiculous pseudo-intellectual game.

The first ornithological book in the modern scientific spirit was published in 1544 by an Englishman, William Turner (1510–1568). He began by attempting to identify the birds named by Aristotle; then he added many original observations. Pierre Belon of Le Mans, France, traveled to the Near East and made many observations; he published books both on fishes (1551) and on birds (1555). At first his work was mainly Aristotelian but he began to see the true nature of the anatomies of various animals and did some excellent work on comparative anatomy. He was the first illustrator of the mammalian nature of the whale and the placenta of its young.

Guillaume Rondelet (1507–1566), a contemporary of Belon, studied the marine animals of the Mediterranean, first to verify Aristotle, then to add many personal observations. His work, published in 1555, is treasured in libraries today and he is still referred to as the first person to illustrate many of the less well-known **invertebrates,** animals without backbones.

RISE OF RATIONALISM

The sixteenth century brought an awakening world. Printing had been developed at the end of the fifteenth century. Travel continued to increase, and the travelers returned bearing collections of plants and animals to be admired, investigated, illustrated, and named. Modern methods of scientific investigation were beginning to be used, and soon the scientific societies and journals for the exchange of ideas would come. Surely science was taking the steps, slowly gathering speed which would increase to the breathtaking pace of discovery in the twentieth century.

The first steppingstones toward expanding the horizons of the descriptive anatomical treatment of all things animal and vegetable were the sciences of physiology and embryology. Speculations on the functions of living parts had been made by the Greeks, but accurate scientific techniques had not been developed for experiment. Too much energy was expended, for example, on arguments about the source of "body heat," the location of the soul, and the origin of life.

William Harvey (1578–1657) awakened England to the new era in science. He also initiated the true physiological approach to anatomy. The students of the Vesalian tradition in Padua no doubt inspired Harvey while he was there from 1597 to 1602; he returned to England ready to move on from the "religious purpose" views of function held by Galen to the more mechanistic purpose of the nature of function. He refused to discuss the philosophical problems of nature and insisted on working with immediate problems that could be dealt with specifically. Thus he established an invaluable attitude among both contemporary and later scholars that influenced not only zoology but chemistry and physics as well. Harvey, nevertheless, revered Galen as the "Father of Physicians" and even credited Galen with knowing about the circulation of the heart and blood vessels which he himself actually discovered.

CONCLUSION

These traditional approaches to the study of animals have occupied a large part of the formal teaching in the years since the Renaissance foundations of the sciences. The growth of knowledge following that period has naturally forced a proliferation of disciplines which, in turn, have narrowed the various fields. Anatomy branched into comparative anatomy, histology, physiology, and embryology. These sciences have been forced to divide again due to sheer accumulation of knowledge and the inability of any one individual to know a subject completely.

Physiology has provided us with an example that shows the historical trend in teaching. The large organs—heart, lungs, stomach, and muscle—once comprised the course content of physiology. But the recent vast growth in knowledge has relegated this approach to an elementary position because scientists have progressed to the cellular level of function and then to the intracellular level. We are now faced with a paradoxical return to the requirements of the early twentieth century when a scientist, or naturalist, was expected to teach chemistry, physics, zoology, and botany, and possibly mathematics. For physiology has pushed back the limits of function to the level of the biochemist's successions of chemical reactions, to the physicist's merry flow of electrons, and to the busy click of the mathematician's computer.

Zoology is a different study today, but it must have its foundations firmly established in the knowledge gained by its pioneers. Of these fascinating people we shall hear more as we deal with the subjects they investigated. We must also examine the new, and sometimes controversial, achievements of our current pioneers in order to appreciate the truly remarkable progress made in recent years. To study only cellular structure would limit us to the same perspective of examining bricks in a brickyard and from them trying to envision a Gothic cathedral or a skyscraper; plainly something would be missing in our appreciation of either of these structures. Conversely, to view the buildings only from a distance, as in a picture, would tell us little of the marvels of construction involved in their making. We must, then, try to look at our subject from both near and far in order to understand the vast scope of modern zoology.

CHAPTER 2

THE ORIGIN OF LIFE

Speculation about the origin of life on earth has taken place throughout man's history and has produced some highly colorful tales as well as some surprisingly astute observations. Once again, we must yield the lead to the ancient Greeks, for they presented a philosophy which, in many respects, strikingly parallels the current concepts on the origin of life.

The Greeks believed that the entire universe was living—that matter was neither created nor destroyed, only changed in form. It was easy for them to conclude that the amorphous slime of the sea could actually generate, under the influence of heat or air, the many living plants and animals that could be observed in it. Anaxagoras, Empedocles, Anaximander, and Democritus all supported theories in which inorganic matter was in some way transformed into living matter.

Democritus fathered the **atomic** concept of matter. He advanced a theory that life was self-created from the natural movement of atoms that met at random and united under the influence of an energizing fire. He and Anaximander both believed that animals had developed over long periods of time, gradually changing until the present-day forms emerged. Certainly these men had a concept of evolution, at least in regard to the origin and development of animals. Democritus even remarked on the survival of various animals as being related to the fitness of their bodily structures, predating Darwin's ideas by more than two thousand years!

SPONTANEOUS GENERATION

A subtle change affected these evolutionary ideas as time passed; when Aristotle chronicled the achievements of science to his day, these concepts had become far more limited in scope. Matter was held to be nonliving, requiring the addition of some active principle or soul (**entelechy**) to form life. He considered that this formation occurred continually, and that the proportion of the influencing factors present—air, fire, and water—determined what sort of living matter would be brought forth. That animals gave birth to like animals was obvious; they passed on "animal heat" to form life. But

"solar heat" could certainly be applied to the generation of life from the slimes.

Aristotle's theories about spontaneous generation influenced scientific thought strongly for the next two thousand years, for with the renewed emphasis on classical learning, his word was considered the primary authority on almost every subject. Even St. Augustine worked out a way in which spontaneous generation could be reconciled with Christian theology so that it became church dogma.

The entire concept of spontaneous generation, consequently, had moved far from the atomic interpretation of Democritus and his contemporaries and had become ever more unrealistic. Often the very few medieval scholars who could read were credited with wizardry. They were seen poring over large dusty books (which contained, perhaps, the contemporary limited knowledge of chemistry) and were observed to mix and stir things that changed in color, form, and odor. No wonder witches and wizards were credited with being able to create living things from their brews.

Travelers returned from faraway places with tall tales, such as those about the oriental melon tree that produced lambs, and the tree that produced ducks. The famous alchemist Paracelsus (1493–1541), no doubt influenced by such tales, produced a recipe for making a little man, or *homunculus,* using human sperm, blood, and some complicated maneuverings.

To fortify these fanciful tales, which may seem ridiculous to us, were the obvious "logical" observations which could be made by everyone. Since nothing was known of bacteria and other microscopic organisms, and their abilities to grow almost anywhere and everywhere, it is not surprising that spontaneous generation was the explanation for the fresh water that soon was teeming with wrigglers, and for other similar natural occurrences. Surely manure gave rise to worms—there were the maggots for all to see. It was not until Francesco Redi (1626–1698?) in Tuscany covered some meat with cheesecloth and observed the flies laying eggs on it, that anyone believed otherwise. In fact, many supposedly learned scholars refused to accept Redi's results.

CONFLICTING THEORIES OF GENERATION

The arguments raged on, with the French Count de Buffon, George LeClerc (1707–1788), and the Scottish minister J. T. Needham (1713–1781) intent on demonstrating that a "vital principle" existed which formed new organisms from the "organic molecules" released by the death of other organisms. By this time, Anton van Leeuwenhoek in Holland (1632–1723) had discovered the world of micro-organisms with the aid of his lenses, to which we will refer later in more detail. Leeuwenhoek became convinced that decay and fermentation resulted from the presence of micro-organisms which invaded clean milk or bouillon, rather than from organisms which arose spontaneously in the liquids.

In 1765, the Italian Abbé Lazzaro Spallanzani set up numerous careful experiments with sealed flasks which he heated; his work showed clearly that the heating kept the liquids clean and fresh. Needham and Buffon immediately claimed that he had killed the "vital principle," so Spallanzani then reopened some flasks which began to show the effects of contamination at once. Still the arguments about the generation of animal life raged on into the nineteenth century. In 1859, Felix Pouchet published lengthy articles which bolstered the "vital principle" and stirred up tremendous controversy.

The French Academy of Sciences then offered a prize to anyone who could perform convincing experiments to end the controversy. Louis Pasteur was awarded the prize in 1862 for a series of brilliant experiments which convinced all but the most stubborn opponents that spontaneous generation was absurd. He contrived ways of showing that the air was filled with organisms by sucking air through cotton and then dropping the cotton in alcohol. The suspended organisms could be seen readily through the microscope. He sealed flasks, filtered, heated, exposed—and proved conclusively that in each case contamination from the air or from objects resulted in spoilage. Pasteur even showed that blood and urine obtained by special methods to avoid any contamination could be kept fresh.

So, finally, spontaneous generation seemed to be a dead issue, disproved and discarded. It is ironic that present-day theories of the primary origin of life are based on a special kind of spontaneous generation. Current theories, more like the concepts of the ancient Greeks than like the medieval lamb tree, nevertheless could be termed "spontaneous generation" provided you are willing to consider as spontaneous the billion or so years that were required for life to occur.

FORMATION OF THE EARTH

The origin of life is bound to the origin of the universe and to the origin of the earth because the nature of the primordial environment would have to provide the working materials for life.

Several theories about the origin of the earth have been developed in the twentieth century. One proposes that the sun, which is observed to throw out enormous flaming tongues that fall back into its surface, was subjected to some extraordinary gravitational pull at a time coincidental with one of these flareups. Perhaps another star passed close enough so that the flaming gaseous mass, instead of falling back into the sun, became detached and whirled into orbit around the sun.

Another more recent theory suggests that the earth and sun are the same age (perhaps 4½ to 5 billion years) and that they were formed from a whirling, hot mass of gases. The sun was formed first, by a concentration of free hydrogen and helium atoms along with smaller quantities of heavier atoms. The remaining gases continued to swirl around the sun in a belt, until other spheres gradually formed from the whirling masses of gas. The glowing hot gases began to separate into various free atoms which then grouped themselves by weight. The heaviest, such as iron and nickel, collected in the center; lighter atoms, such as silicon and aluminum, formed a middle layer; and the lightest—hydrogen, helium, nitrogen, carbon, and oxygen—formed the outer layers. As the new solar system cooled, the motion of the individual atoms was slowed so that they formed attachments to one another instead of colliding and bouncing away in individual random flights.

A recent theory, by the British astronomer Fred Hoyle, favors a "cold" formation caused by huge clouds of dust. When the dust began to whirl, the particles generated heat by friction. This increased molecular activity so that gases were formed. Whether formation was "hot" or "cold," heat would have been generated and the same changes would then follow.

The changes that then came about varied according to the individual elements, for each has its own specific temperature and pressure at which it exists in the three different states of matter: solid, liquid, and gas. We have observed in our own experience that iron is solid in our normal environment and we know that only when tremendous heat is applied, as it is in steel mills, will iron become liquid. Under special conditions of extreme heat, iron can even be made to vaporize. Physicists tell us that increased pressure causes an increase in heat because more atoms collide in a given space. The heavy iron which settled in the center of the earth was subjected to pressure and not exposed to surface cooling so the core of the earth has theoretically remained hot and molten.

Scientists visualize at this stage a crust forming as some of the gases cooled to liquid state and some of the liquids cooled to solids in the new sphere. The forms of water are familiar—steam (a gas) from a teakettle condenses to droplets of liquid on striking a cool surface; the liquid turns into ice, a solid, if it is cooled further. These changes occurred on the young earth.

The characteristics of this newborn earth are very important because they determine whether or not life could have originated from the raw materials available. Unfortunately, we do not have any direct positive evidence as to what the primeval earth and its "atmosphere" contained. By examining the other planets with the modern equipment and methods of astronomers it has been concluded, for example, that ammonia and methane (which are gases on earth) and water are present on Jupiter. They are, however, in a frozen, solid state due to Jupiter's great distance from the sun. This probably has prevented chemical reactions from occurring there, and makes the possibility of life as we know it rather remote. Analysis of the spectrum of the light from the sun tells us what elements are present there. Since the earth supposedly originated from the sun or with it, the elements should be the same in both bodies.

DEVELOPMENT OF LIFE

Speculation on the origin of life has taken two main directions. One theory is that living matter came as some sort of space-traveling "seed" to furnish the vital essence needed for all living things. The other is that life arose from the inanimate matter of the earth. These conflicting opinions existed before Aristotle's time and still continue.

Only recently a flurry of discussion developed when it was claimed that bacterial material had been found in meteorites. It seems much more plausible that contamination of the meteorites occurred in the earth's atmosphere than that the bacteria rode the fragments through heat, cold, and radiation in deep space and arrived undamaged.

The idea of life arriving from elsewhere has attracted, in the past, quite a following, particularly during the first half of the last century. Justus von Liebig, the German chemist, wrote in 1861 that "the atmosphere of celestial bodies as well as of whirling cosmic nebulae can be regarded as the timeless sanctuary of animate forms, the eternal plantations of organic germs." H. Richter, a German biologist, and his countryman, the famous physiologist and physicist Hermann von Helmholtz, both favored the idea of cosmic traveling germs.

The Swedish chemist Svante Arrhenius in the early 1900s became an advocate of the theory of universal life and produced some complex calculations on the travel of minute particles in space under the force of light rays. Arrhenius was a highly respected physical chemist, and his ideas had many believers. He suggested that germs could be transferred from one planet to another by passing through trails of particles that were pulling away from the planets like comet tails.

In the 1930s, Russian scientists began actively to investigate the possibilities of the origin of life on the earth. A. I. Oparin of the Biochemical Institute in Moscow first wrote a short booklet in 1923 on the subject and, in 1936, wrote *The Origin of Life* which is considered the definitive work. It was translated into English in 1938, and reissued in 1953. Much of the current teaching and research is based on Oparin's ideas, which have stood up well under the pressures of advancement in biochemical knowledge.

ATOMS, MOLECULES, AND ELEMENTS

We have seen that the outermost layer of the slowly cooling earth was composed primarily of the lighter atoms: hydrogen, helium, oxygen, nitrogen, and carbon. The form in which these occurred, however, has been disagreed upon by authorities. Some elementary knowledge of the properties of these atoms is essential for an understanding of the theoretical development of life.

Atoms. Atoms are the smallest complete units of elements. Ninety-two different elements occur naturally, several more have been formed artificially in nuclear reactors. Each atom has a nucleus which

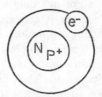

hydrogen atom

Fig. 1

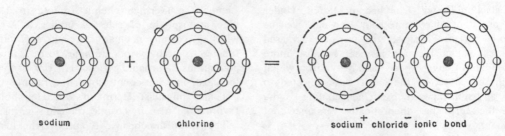

sodium   chlorine   sodium $^+$ chloride $^-$ ionic bond

Fig. 2

contains a definite number of **protons** (positively charged particles) and **neutrons** (no electric charge). Usually, the protons in the nucleus are surrounded by an equal number of **electrons** (negatively charged particles). Each element has a specific **atomic number,** which is equal to the number of protons in the nucleus. The **atomic weight** equals the number of protons plus the number of neutrons.

Some elements are composed of more than one form of atom; that is, the atoms differ in the number of neutrons, although they all react the same chemically because they have the same charges. These are known as **isotopes.** There are, for example, three kinds of hydrogen and five kinds of carbon isotopes. Counting all known isotopes the elements then number more than 1600.

The electrons travel around the nucleus of an atom in certain clearly defined orbits. Atoms of different elements have from one to seven shells, named K, L, M, N, O, P, and Q, in which electrons move. The pathways or orbits of the electrons in the shells determine the action of the atom.

The outermost ring of the atom is the one in which chemical activity will take place if it is possible. The innermost (K) shell, has room for two electrons. The second (L) and third (M) shell each have room for eight. If the outside shell of an atom has the full number of electrons, it is inert and does not combine with other atoms. Helium and neon are inert gases under normal conditions. If the outer shell is not full, the atom will be chemically active.

Forming Simple Molecules—Inorganic Synthesis. Each atom is, in a sense, going to try to fill its own outside shell by capturing an electron from some other atom. If the atom has only a few of the total complement of electrons it can give up its outside shell and, like peeling an onion, expose the next shell inside, which has a complete complement of electrons, and thus become stable. Then the atom that has most of its outside shell filled will pick up the "shed" electrons of the other to fill out its own shell. The "electron losers" are said to be **electropositive elements** because when electrons are lost the positive charges in the nucleus exceed the negative

charges of the electrons in orbit. Conversely, in the "electron catchers," the atoms will have more negative charges in the shells than the nucleus has positive charges, so they will be **electronegative.** The electrically charged atoms are called **ions.**

After this electron transfer has taken place, the gainer and the loser do not separate and go their own ways, for we find that the unlike charges attract the two atoms and an **ionic bond** is created. A good example of this type of bond is found in our common table salt, sodium chloride. Sodium has three shells, with only one electron in the outer shell. Chlorine has three shells also, but it has seven electrons in the outer shell. Sodium gives up one electron, losing its third shell, and chlorine takes the electron to complete its third shell. The two atoms are held together by their charges; the atom of sodium and the atom of chlorine form what is called a **compound,** sodium chloride (Fig. 2).

What of those atoms that have four electrons in their outer shell? Carbon is a good example of this; it usually solves the problem by sharing its electrons without losing them. This gives carbon some properties that are important to living systems, as we shall soon see.

Other atoms share electrons also. Hydrogen, the lightest of all elements, has only one electron in a single shell. In order to complete this first shell, which requires only two electrons you remember, two hydrogen atoms share their electrons and thus form a **molecule** of hydrogen gas. Each atom shares the electron of the other, forming what is called a **covalent bond.** Actually, they are overlapping the orbits of the shared electrons, for it is not a passive

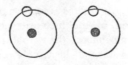

hydrogen atoms   hydrogen molecule
                         covalent bond

Fig. 3

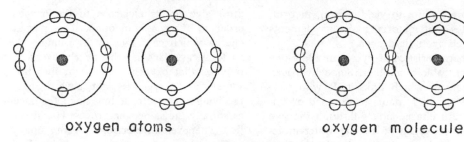

oxygen atoms   oxygen molecule

Fig. 4

sharing. You have no doubt seen two children, each with a jumping rope, face each other and co-ordinate the rope turns so that the orbits overlap. If there were a bead at the top of the loop of each rope, you could visualize that bead as being the shared electron and each child would be the nucleus of one atom. See Fig. 3.

Oxygen also forms molecules in this way: each atom has two shells, with six electrons in its outer shell. Each atom shares two electrons with another so that a molecule is formed with shared electrons holding them together. See Fig. 4.

Mixed molecules also may be formed by such sharing. A molecule of water is formed when an oxygen atom with six electrons lies between two hydrogen atoms. The oxygen shares two electrons, one from each hydrogen atom. Methane is formed by a carbon atom, which has four electrons in its second outer shell, sharing electrons from four hydrogen atoms. Ammonia results when nitrogen shares three electrons from hydrogen atoms with the five electrons it has in its third outer shell. Our illustrations of the bonds show only the active (outer) ring of each atom. The formulas can be written by adding up the atoms, for example H_2O, NH_3, or CH_4. In order to represent more accurately the arrangement of the atoms, structural formulas are often used:

Molecule	Water	Ammonia	Methane
Molecular Formula	H_2O	NH_3	CH_4
Structural Formula	H—O—H	H—N—H with H below	H—C—H with H above and H below

Fig. 5

Atoms are too small to be seen with a microscope, so this information has been secured by indirect experimentation. The electron microscope is now ap-

proaching atomic-sized particles; however, individual atoms are not yet visible.

Properties of Simple Molecules. It is obvious that our newly made "earth," with its superheated atoms charging around colliding with each other, had to undergo some simple chemical changes as soon as the atoms cooled enough to stay in contact with each other. Activity of atoms always increases with an increase in heat, and under extreme heat no bonds will form, or bonds will break immediately after they do form.

From the properties we have discussed above it can be seen that molecules of ammonia (NH_3), water (H_2O), and methane (CH_4) could have been present along with molecules of hydrogen (H_2), and some oxygen (O_2) in our primitive environment.

Oxygen is a very active element and can unite with many of the other elements to form <u>oxides.</u> We are familiar with iron and how quickly it "rusts"; it is forming iron oxide by the union of oxygen from the air with the iron.

Because of the ready ability of oxygen to combine with other elements, the primitive environment would not have had much free oxygen after the atoms cooled sufficiently to begin uniting with each other. One of the most common oxides would be oxide of hydrogen, which we recognize as water.

The earth's first atmosphere (resembling the sun) may have been a blanket of hydrogen gas, alone or with a few other elements present. Hydrogen combined with other elements such as nitrogen and carbon to form ammonia and methane. These substances are poisonous to life as we know it, so that other events must have occurred to create a habitable environment. Some authorities feel that many of the lightweight hydrogen atoms simply wandered off into space when their bouncing about took them too far beyond the weak gravitational pull exerted on the light atoms by the large mass.

A different theory of what happened to some of the hydrogen may be found in the experimental research done for the hydrogen bomb. Under certain conditions, hydrogen converts to helium. Helium then fuses to form an isotope of beryllium, which interacts with more helium to yield carbon. In principle, this indicates that all of the various elements could have evolved from hydrogen.

The peculiar characteristics of hydrogen allow one molecule of water to join another. Since hydrogen has only one electron, which links itself with oxygen, the rest of its shell remains unoccupied or unguarded. The proton charge pokes through the gap and hooks on to the oxygen of another water molecule, eventually gathering enough molecules to form droplets. (This extra "hook" from the hydrogen nucleus is a special bond that is unique to hydrogen.)

And so the rains came. The earth had finally cooled sufficiently to reach the temperature at which hydrogen united with oxygen to form water molecules, water vapor condensed into water, and it began to pour. Since all of the water up to this time was in vapor form, a huge cloud layer several miles thick, which helped the cooling process, must have surrounded the earth. Methane, ammonia, and other gases can be dissolved in water when it is cool enough and the rains brought these substances down with them.

The crust of the earth was probably thin, with the fiery hot molten rock not far below its surface. The crust had already been pushed into wrinkles and ridges by the contractions of cooling and by the bursting-out of the molten inner material from below. Now the rains roared down the slopes on the crust and collected in low spots and in basins to form the ancient seas of which we see the evidence.

The land was not yet cool enough to allow the seas to remain; like a giant soup kettle boiling away, it turned them back into vapor. How long did it rain? Well, it has never entirely stopped raining, for the same process, on a more limited scale to be sure, still goes on. Judging from the intensity of some of our present storms, the weather must have been cataclysmic for millions of years.

In addition to the gases brought down in the rainfall, the torrents of rain eroded the earth's surface and carried to the ocean various mineral elements and compounds. We find our soup kettle analogy apt once more because the kettle now contains such a great variety of substances. The most important is, of course, water. Nothing lives without water; it composes 90 per cent or more of all living things; it is also the carrier for many other substances because so many things will dissolve in it.

With the dissipation of all the free oxygen and free hydrogen that were present at first, water would be the best source of these two atoms if they are needed again. As it happens, hydrogen and oxygen are vital parts in the energy systems of **cells,** which are the basic units of living organisms.

In the process of oxidation, some kind of atom must give up its electrons to the oxygen atom in order to form a compound. This process releases energy which cells employ in running the factory, so to speak. This is why nearly every living thing requires that oxygen be brought into its system in some way. Oxygen is not always the agent for receiving the electron, but since the mechanism is essentially the same, the process is still called oxidation whenever an electron is given up.

The release of energy in oxidation is due to the position of the electron before and after it unites chemically. In the atom, the electron is far out from the nucleus; it is said that some kind of energy was used to put it out there and so it has stored, or **potential,** energy in it. You might compare this with a rock sitting at the top of a hill—it has the energy stored in it which could be felt if it fell on you below. **Kinetic** energy is the energy expended during the motion of the rock while it falls. Perhaps a better analogy would be the potential energy of the water at the top of a waterfall compared to the kinetic energy being expended as the water falls, striking air and rocks as it goes. It has less energy the nearer it gets to the bottom.

The electron which is to be given to the oxygen atom starts far out in its orbit and "falls" toward its center of gravity, the nucleus, when it combines with the oxygen. Thus, energy is released.

Reduction, the gaining of an electron rather than its loss, is a necessary partner to oxidation. If an electron is taken from one atom by oxidation, it must be received by another atom, because electrons cannot exist in a free state. The electron must be accepted by an oxygen atom or atom of another element, which is said to have been reduced—it has added a negative charge. This exchange is known as an **oxidation-reduction reaction.**

Without getting too deeply involved in the chemistry and physics of living systems, we can conclude then that water, with its component parts, hydrogen and oxygen, is necessary to the energy system. We can also conclude that because of the nature of our primitive environment water existed in great quantity if the conditions were right for forming living systems.

ORGANIC COMPOUNDS

Organic compounds are compounds that are formed with carbon. At first glance, this might seem a limited group, but that is not the case at all. As we have mentioned previously, carbon has some very interesting talents based on those four electrons whirling around in its outer shell.

Probably the first activity in which the carbon atoms participated was in the formation of **carbides**

—compounds formed when carbon atoms join the atoms of other elements. Calcium carbide is an example of one of these compounds found deep in the earth, where they were carried when the heavier elements, such as iron, nickel, and cobalt, sank toward the center early in the earth's existence. See Fig. 6.

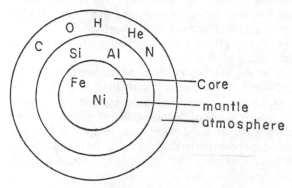

Fig. 6. Location of Some Elements in and around the Earth

Sometimes the carbides did not stay in the inner regions of the earth, but were brought to the surface when molten volcanic material broke through the outer crust. Carbides immediately react with water when it is present; the carbon atoms link with one of the two hydrogens in a water molecule. For example, when calcium carbide (CaC_2) comes in contact with water (H_2O), calcium hydroxide ($Ca(OH)_2$) and the gas acetylene (C_2H_2) are formed:

$$CaC_2 + 2H_2O \longrightarrow Ca(OH)_2 + C_2H_2 \uparrow$$

| calcium carbide | water | calcium hydroxide | acetylene gas |

The calcium has taken the oxygen and one of the hydrogens from the water molecules; this (OH) is called the hydroxyl group. The carbon has taken the remaining hydrogen to form the gas acetylene.

The formation of acetylene shows us a very important ability of carbon, that of linking with hydrogen to form compounds known as hydrocarbons. The linkage between the carbon and hydrogen atoms can readily be exchanged for other atoms. For example, we mentioned earlier methane (CH_4) that formed in the first chemical reactions we encountered. Any one of the hydrogens in methane may be replaced, and two, three, or all four may be replaced by other atoms.

Another special talent of carbon is that of joining with other carbon atoms so they appear to be in chains, like so many children forming a line by holding hands. Carbon can form rings just like the children form a circle. Since carbon has four electrons to share, holding bonds on either side still leaves two

bonds free for some other atoms to hang on to. Sometimes carbon holds onto its neighboring carbon on one side with a **double bond** (two bonds), and the neighbor on the other side with a single bond. This leaves only one bond open for another attachment to each carbon.

Benzene, C_6H_6

The compound benzene is a ring of six carbons held in place by alternate single and double bonds. Thus, we have many compounds that can be formed just because of the unique ability of carbon to form bonds with other carbons, and to transfer bonds from hydrogen to other atoms.

We mentioned previously the (OH) that was left when a hydrogen was split from water. This, you remember, is called a hydroxyl group and acts as one unit, so that a hydroxyl may replace a hydrogen held by a carbon. In this way the compounds called **alcohols** are formed. The simplest begins with methane (CH_4),

By replacing one H with an OH we have CH_3OH, or methyl alcohol:

If there are more carbons linked together in a chain, a different alcohol will be formed. Also, more than one hydrogen may be replaced by a hydroxyl. The location of the carbon in the chain that does the replacing makes a difference in what kind of compound is formed. For this reason, organic compounds are diagrammed as structural formulas rather than by listing the number of C's, H's, OH's, and the other atoms participating. Clearly, carbon has the ability to form a wide variety of complex molecular structures.

Why are the special abilities of carbon important to us in studying zoology? The answer is that all living things are composed of compounds that contain carbon, and all the compounds produced by living things contain carbon. The properties of being able to join other atoms by one, two, three, or four bonds, and of being able to join with other carbons in long chains or in rings, form the basis for the actual physical structure of plants and animals. These substances are called **organic compounds.**

Inorganic Compounds. Organic compounds differ from the chemical compounds formed by chemical interaction between the elements that are found in the earth and the atmosphere. Such substances as water, metals, and salts occur without the presence of life and are known as **inorganic** compounds. Inorganic substances are used by living organisms, however, who add them to the organic structure as they are needed. Later we shall see the importance of some of the inorganic compounds.

Origin of Organic Compounds. In earlier times scientists thought that it was impossible to form organic compounds in the laboratory because they required some "vital substance" present only in living things. Years ago this was proved wrong; organic compounds can be made (synthesized) in the laboratory. Some of the processes are relatively simple, although others are extremely complicated.

After it was established that organic compounds could be synthesized in the laboratory, scientists began to wonder whether similar processes might not have occurred in the primitive seas. The seas were warm: reactions could have been stimulated by the heat; and dissolved materials were in plentiful supply. We have seen that the carbon, hydrogen, and oxygen necessary to form basic organic compounds were present. We have also seen that methane and acetylene could have been formed, as well as water and ammonia. From the methane molecule, simply by the replacement of hydrogens at various bond locations, more complex organic substances like alcohols could also have been created.

CARBOHYDRATES

Much of our present-day food supply is based upon compounds formed of carbon, hydrogen, and oxygen. **Carbohydrates,** the starches and sugars, are such compounds. Starches are present in wheat, corn, and rice, for example; sugars are found in sugar cane, sugar beets, fruits, and other foods.

The term "carbohydrate" refers to the addition of water to carbon, and the general formula shows a proportion of one carbon to two hydrogens to one oxygen. The structure differs from other hydrates so the term **saccharide** is replacing the word carbohydrate in many texts. The sugars are named on the basis of the number of carbons present; hexose sugars ($C_6H_{12}O_6$) are the most common, but five-carbon sugars, pentoses, are very important in the living cell, as we shall see later.

Starches are sugars from which water has been removed chemically, so that they may be concentrated for storage. The sugar molecules may form in straight carbon chains or linked rings. This enables them to build very large molecules that are called **polysaccharides,** which can form such large structures as the outer coverings of certain animals like crayfish.

FATS

Fats are also important substances formed from carbon, hydrogen, and oxygen. They become part of animal bodies, and as such are energy-producing foods, since they provide twice as many calories as carbohydrates of equal weight. Fats are formed when one of the alcohols, glycerol, combines with a fatty acid to form an **ester,** the "salt" of a fatty acid. Glycerol and fatty acids are both formed from digested sugars.

Fatty acids are composed of long chains of even numbers of carbon atoms, each one holding on to two hydrogen atoms. At the end of the chain is a group of atoms that act as a single atom, the **carboxyl group (COOH).** Every organic acid ends in a carboxyl group, no matter what the rest of the structure may contain. Fats contain very little oxygen compared with the sugars; beef fat, for example, has a formula of $C_{57}H_{110}O_6$.

THE ROLE OF NITROGEN

We have said little about the importance of nitrogen up to this point. If nitrogen is added to the three necessary elements—carbon, hydrogen, and oxygen—it is possible to build **proteins,** which are the basic structural material of animal bodies, and also occur in plants. Two other less well-known substances are purines and pyrimidines, which, as we shall see later, form part of the genetic code molecules in the nucleus of the cell.

Origin of Nitrogen Compounds. If it is possible that life originated in the primitive seas, it is necessary that we examine the way nitrogen might have been brought into the organic compounds to eventually form proteins and the other important nitrogen-bearing substances.

Nitrogen atoms were no doubt present in the original atmosphere, and may have combined directly with hydrogen to form ammonia. It is also possible that nitrogen combined with carbides to form compounds called cyanamides ($CNNH_2$), which are used today as important fertilizers.

When cyanamides are brought into contact with water, ammonia is given off. For example, calcium cyanamide plus water gives ammonia and calcium carbonate (lime). If, as some authorities believe, carbides were carried into the center of the earth and then forced out in volcanic eruptions they would have combined with the nitrogen gas in the atmosphere to form cyanamides:

$$CaC_2 + N_2 \longrightarrow CaNCN + C.$$

calcium   nitrogen   calcium   carbon
carbide   cyanamide

When these in turn came into contact with rain or sea water ammonia would have been released:

$$CaNCN + 3H_2O \longrightarrow CaCO_3 + 2NH_3.$$

calcium   water   calcium   ammonia
cyanamide   carbonate   gas

Ammonia can be formed in yet another way that might have occurred in the primitive environment. Nitrogen unites directly with metals to form nitrides, just as carbon combines to form carbides. When nitrides are exposed to water they, too, give off ammonia.

Importance of Ammonia. Ammonia may not seem like a compound of great importance to us because we use it casually for cleaning tasks. But the presence of large quantities of ammonia in the primitive environment is significant because it is essential for the formation of **amino acids.** Amino acids link together to form **peptides;** peptides link into chains called **polypeptides;** and long chains of polypeptides form **proteins.**

The specific formula of each of these is not important, but it is important to understand the structural characteristics of the various compounds to see how this sequence of events can occur.

Let us start with the gas methane (CH_4), for example:

We know that any or all of these hydrogens may be replaced by other atoms or groups of atoms acting as one. Suppose, then, that we replace the right-hand hydrogen with the organic acid carboxyl group (COOH):

This is now an organic acid, acetic acid, which is present in vinegar. The generalized formula for any organic acid is RCOOH; here R is CH_3.

If ammonia now comes into proper contact with this acid molecule, an amino group (NH_2) will replace the left-hand hydrogen to form the simplest amino acid, glycine:

More complex amino acids can be formed by replacing one more of the two remaining original hydrogens. The group or chain of groups that replace hydrogen at the top or bottom position will be known as a **side chain.** The side chain will determine the kind of compound formed, but it will not affect the primary chemical reactivity of the amino acid, as we shall see shortly. Therefore we can represent the side chain as the letter R.

The Russian Oparin has speculated that in the primitive atmosphere all of these reactions could have occurred spontaneously. The American chemist Harold C. Urey elaborated on the theory, and one of his students, Stanley Miller, succeeded in synthesizing amino acid, using electrical discharges for energy, in 1953. Others have since done this using radiation for energy, and have also been able to form a variety of organic compounds such as fatty acids, purines, and pyrimidines.

Peptide Linkages. Amino acids have a very interesting property: they are positively charged at the amino end (NH_2), and negatively charged at the carboxyl end (COOH). Depending on their surroundings, they may act as an acid or an alkali (base). This also means that the end of one amino acid is attracted to the beginning of the next, since unlike charges attract, and they will react by splitting out a molecule of water, linking the remains of the acids into peptides. This is called a **peptide bond,** and is an example of a bond known as an **anhydro bond** since it removes water. Many of the linkages in animals are composed of such bonds. Conversely, digestion of food often involves putting the water back in, to split up the linkages:

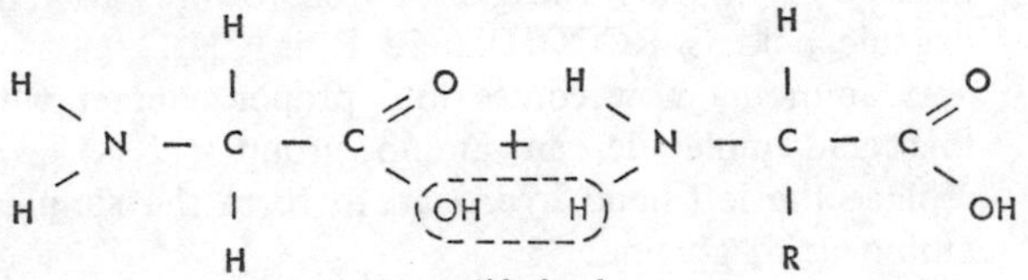

peptide bond

This bond can also be remembered as the COHN bond, since those are the substances joined together when the water is removed.

FORMATION OF PROTEINS

Peptides may join together in long chains to form polypeptides. These may in turn be joined to form proteins, which are very large molecules. The chains fold back on themselves and it is here that the importance of the side chains of the amino acids, the R groups, can be seen.

The side chains may contain atoms other than the basic carbon-hydrogen-oxygen-nitrogen structure. They may form bonds between themselves that keep the protein stabilized in its twisted form. Sulfur linked with hydrogen in the SH form (sulfhydryl) will, by removal of hydrogen through oxidation, form a strong S-S bond, called a **disulfide bond.**

The unique hydrogen bond that we spoke of earlier will also act to stabilize the protein structure. The remaining H of the amino group will be attracted to the O that remains of the carboxyl group of nearby peptides. In this way proteins become the body-building blocks.

Proteins as Enzymes. Proteins also perform a very important function in addition to building cells and tissues in the body. They act as **enzymes,** the catalysts of the living world. A **catalyst** is a substance that speeds up a process without being chemically altered itself at the conclusion of the reaction.

PURINES AND PYRIMIDINES

Purines and pyrimidines are organic structures that are based upon a ring of carbons similar to the benzene ring. The pyrimidine ring is the more similar, for two of the six carbons in it are replaced with nitrogen. Since each carbon and nitrogen may hold different atoms by their non-ring bonds, different pyrimidines may be formed. Cytosine, uracil, and thymine are three pyrimidines that help to form the

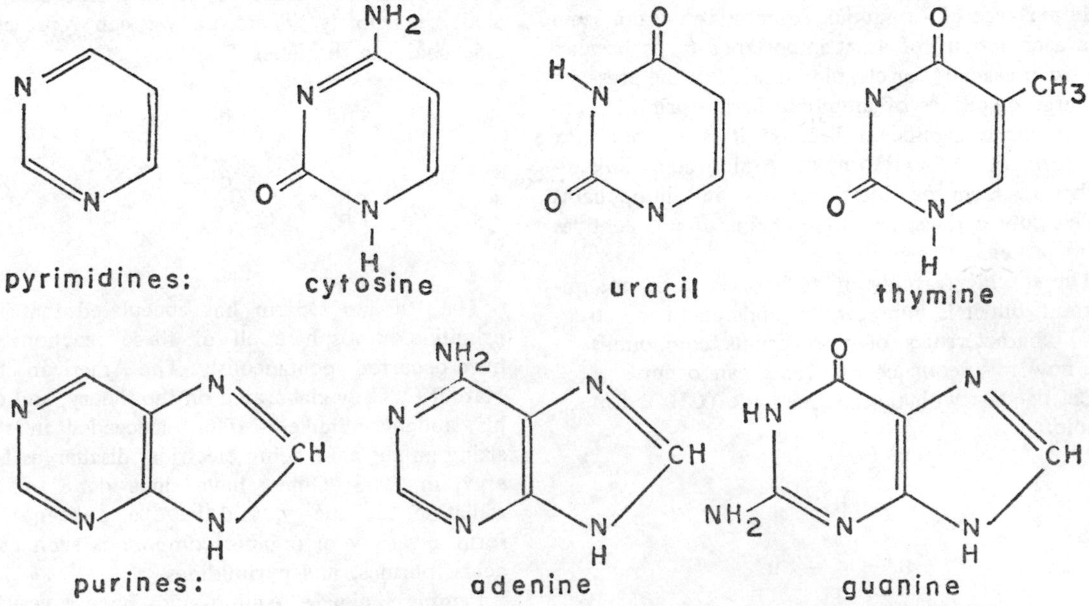

pyrimidines: cytosine   uracil   thymine

purines: adenine   guanine

Fig. 7. Some Typical Pyrimidines and Purines

genetic material in cells so that they can reproduce themselves.

Purines are composed of two rings, a six-sided ring linked to a five-sided ring. Here again several of the carbon positions on the rings have been taken by nitrogen, and different atoms may be held by the side bonds. The purines that are important in genetics are adenine and guanine. (All of these substances are called **nitrogenous bases.**) Purines and pyrimidines are able to link to each other by weak hydrogen bonds which can be separated easily; this characteristic is very important in allowing them to function in reproduction. (All the possible hydrogens are not usually shown on structural formulas.)

PHOSPHORUS AND PHOSPHATES

We have not previously discussed phosphorus, one of the elements that is thought to have been present at the formation of the earth. Phosphorus combines with oxygen to form phosphates. Because of the arrangement of their electrons, phosphates are able to carry energy in living organisms, as we shall see when we discuss organic maintenance in Chapter 7. The actual way in which the phosphates do this is a subject of controversy and involves the complex field known as thermodynamics, into which we need not venture.

NUCLEOSIDES AND NUCLEOTIDES

When one of the nitrogen bases encounters one of the pentose sugars (five carbons), either ribose or deoxyribose, the base becomes linked by one of its nitrogens to the sugar:

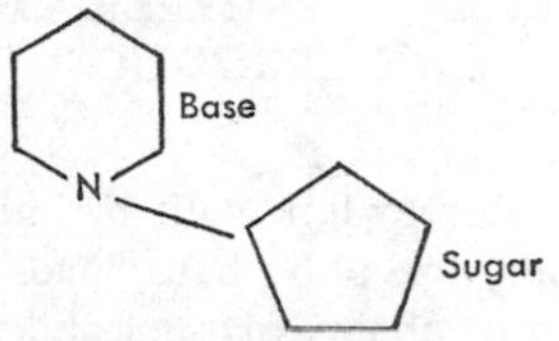

This base-sugar combination is called a **nucleoside.** When the sugar part of a nucleoside comes in contact with phosphoric acid (H_3PO_4), the sugar picks up the phosphate by splitting out a water molecule in an anhydro bond. Additional phosphates may link to that single phosphate, forming mono-, di-, or triphosphates. A nucleoside with phosphate added to it is known as a **nucleotide.**

Nucleic Acids. Nucleotides may be linked together in chains, as when the phosphate group attaches to two separate sugar-base molecules at once. Long sequences of linked nucleotides are called **nucleic acids.** Because of the many exposed bonds, par-

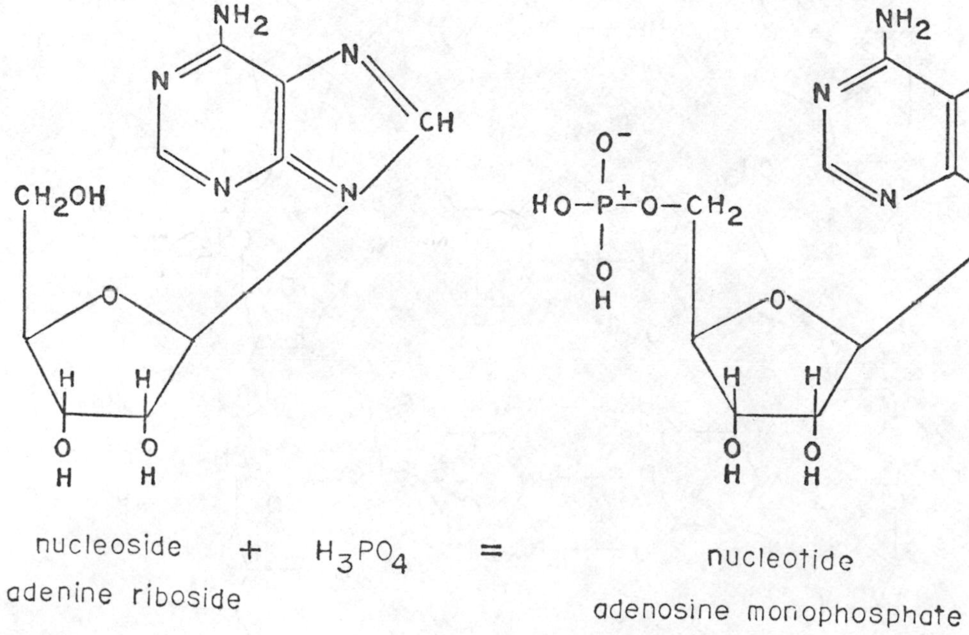

nucleoside   + H_3PO_4 =   nucleotide

adenine riboside   adenosine monophosphate

Fig. 8

ticularly of hydrogen, in proteins and in nucleic acids, these very large molecules easily become linked to one another. Since some of the bonds are weak, however, the molecules may be pulled apart easily under certain stimuli. It is the structural characteristics of these molecules that lead to the formation of life.

ABILITY TO REPLICATE OR REPRODUCE

Purines are able to link with pyrimidines through hydrogen bonds; the phosphate bonds that link nucleosides to form chains of nucleotides are stronger than the hydrogen bonds. The nucleic acids tend to form long strands that are coiled or twisted. These strands, because of the matching up of purines and pyrimidines in adjacent strands, also tend to pair so that they form double-stranded nucleic acids curled into the shape of a helix, like a twisted ladder. If the strands of a pair are pulled apart by some force, and the nucleic acid is in a solution containing bases, sugars, and phosphates, the various bonds will attract these parts so that they line up along the strand in the proper order to form a duplicate strand. The nucleic acid has thus reproduced or built a replica of itself.

We have traced the way in which simple elements and compounds could have become linked together to form many of the various substances that we know comprise and nourish living organisms in the modern world. It is because the primitive seas contained the simple substances in quantity that bigger molecules could have been formed as they drifted around in contact with one another. There need be nothing operating in this system other than the usual bonding mechanisms that occur between atoms, a good supply of the atoms, and a supply of energy to make the reactions occur.

Of the many possible complex molecules that may have been formed in the continual swapping process that went on, several that are essential to present-day living systems could easily have been formed. The nucleic acids, **deoxyribonucleic acid (DNA)** and **ribonucleic acid (RNA)** have been shown to be the threads of life that duplicate and enable cells to form new cells, and they direct the formation of protein necessary for living systems. More will be said about these activities in the chapters on genes and on cell metabolism.

PRIMORDIAL ENERGY

Physical properties of the primitive environment aided in this process by providing ample energy for

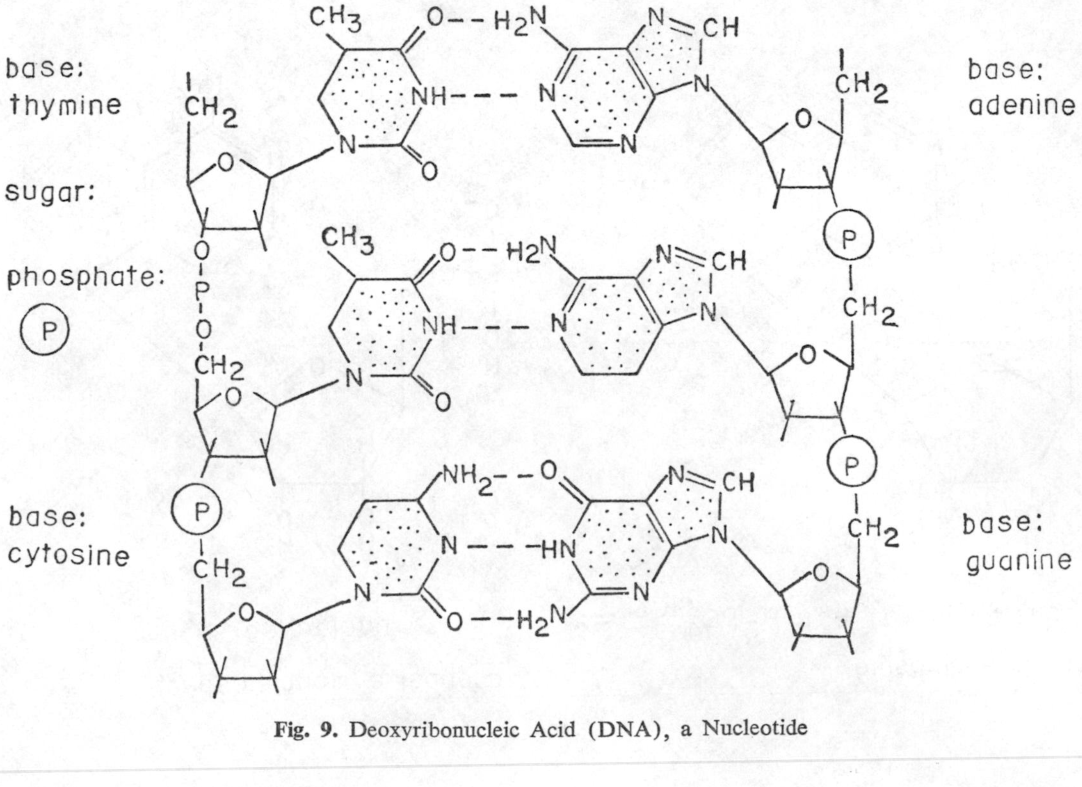

Fig. 9. Deoxyribonucleic Acid (DNA), a Nucleotide

the activity. Ultraviolet radiation (the ray that causes sunburn even on a hazy day) was much stronger then than it is now. The first hydrogen layer around the earth had been dissipated. The only cover between the earth and the sun was a cloud layer of water vapor that had gases, such as methane and ammonia, mixed with it.

At present, a great blanket of oxygen (O_2) fills the atmosphere around the earth; farther out, oxygen occurs in the form of ozone (O_3). It is thought that the present oxygen has been formed since the advent of living systems (plants) and did not exist in the primordial world. Ozone acts as a filter, removing a large proportion of the ultraviolet light so that high-energy radiation is greatly reduced.

It has been demonstrated under laboratory conditions that ultraviolet radiation can provide energy to tie small molecules together, forming larger ones. Amino acids can be combined to form proteins by subjecting them to ultraviolet radiation and hot steam, partially duplicating a process which might have taken place in the primitive environment. To be sure, the proteins formed in the laboratory so far are not living organisms.

Some researchers previously thought that it would be possible to create living matter with such an experiment. Most workers feel that the origin of "life" was a much more gradual process, built of many complex organic reactions that required millennia for development. It is a system of interactions which is too delicately balanced to have been produced abruptly out of a batch of a few reagents. Eventually, a satisfactory sequence of chemical reactions may be found to produce a simple "living" system. The shapes of the particles formed in the present experiments are intriguingly similar to those of some bacteria. In 1965 Sol Spiegelman at the University of Illinois succeeded in "creating" a self-propagating virus from chemicals mixed with an enzyme called replicase.

Lightning is another source of energy that probably helped unite some molecules. The cycle of rains no doubt produced, along with the deluges of water, some awesome electrical displays. Lightning forms ozone by combining oxygen molecules and probably contributed to building up the ozone layer around the earth. With ultraviolet radiation and lightning, the primitive "soup" had a good supply of energy to keep it stirring.

PHYSICAL PROPERTIES OF MATTER

The physical properties of particles are pertinent because they, too, were responsible for assisting new complex reactions to take place.

When particles are placed in liquid, they may dissolve and become part of the solution. Sodium chloride, which is held together by an ionic bond, will dissolve in water to form a solution. In the solution, the sodium and chloride ions separate, but they are still attracted to each other, and consequently, they bounce around through the medium. All molecules in a liquid are constantly in motion, although the individual atoms are too small for us to be able to observe their dances, of course.

If the particles placed in a liquid are not soluble, any of several things may happen, depending on the size of the particles. If they are too large, the molecules dashing around in the solution will not be able to push the particles into the dance, so the particles will simply gravitate to the bottom and settle out.

If the particles are more finely divided, the colliding molecules have more exposed surfaces to bump into. Since the mass of each particle is small, they can be pushed and pulled about by the movement of the molecules and, thus, kept in suspension. When the particles are suspended in this way they tend to act as if they were sticky, causing molecules to **adsorb** on their surfaces. The particles, by carrying along a coating of various molecules, increase the chances of different molecules coming together, and so act as catalysts in uniting larger molecules, which are called colloids. See Fig. 10.

particles

colloids

ions

Fig. 10

Colloids. When colloids stick together, they tend to collect into larger particles and may settle out. Because of their increased mass, the movement of the dashing molecules may not be able to keep the particles moving and they will sink or coagulate. The colloids that adsorb water molecules are able to forestall coagulation by holding a water "skin" about themselves so that other particles will not adhere to them.

Colloids may group together in chains, branches, or balls. Proteins are colloids, and various shapes are formed by different proteins. In animals, for example, fibrous proteins take such forms as muscle strands and part of the bone.

It must be emphasized that the complex protein molecules are not static or stable throughout. They are continually transferring electrons, thus producing chemical changes; parts of the chains may break down and be lost while other parts are building. This characteristic gives us a preview of the metabolic activity that is a feature of living things.

Certain proteins also show the special talent of being coagulated part of the time and uncoagulated at other times; these are the **gel** and **sol** states respectively. Gelatin demonstrates a protein that can be a sol when warm and a gel when cool. Egg white, on the other hand, is a sol in its normal state; when it is heated, however, it turns into a permanent gel.

These states apparently depend on the effect of the activity of the electron exchanges on the long, complex protein molecules. Although the gel state is one of coagulation, all the molecules do not necessarily sink to the bottom. If the molecules are in long strands, they may form a meshwork in the liquid and not sink. Water is trapped in the strands and cannot move around to provide the activity that keeps the solution in motion; so it stiffens into a gel. The complex electron exchanges make it possible for a protein to be in sol in one area while it is in gel elsewhere.

Molecules are sticky; cohesion between molecules tends to hold them closely together. Surface tension keeps water gathered in droplets rather than spread out in sheets. These physical characteristics are true of colloids. Colloids tend to gather in clusters and coagulate or settle out, but the adsorbed molecules that stick to the clusters, and other forces, tend to keep them suspended in a kind of compromise. The resulting droplet of material surrounded by its water molecule "skin" is called a **coacervate.**

Was the primitive cell simply a kind of coacervate? We can only speculate about this. Protein coacervates keep an energy supply going by the flow of electrons. This process also breaks down one part of the protein to create another. By gathering new molecules, the coacervate can "feed" and "grow." It will be stable only so long as it continues to bring in molecules for energy and building. If the supply of molecules is cut off, the coacervate breaks apart completely.

All of these physicochemical qualities are seen in living systems but they are not in themselves unique to life. They are qualities that were present in matter before life developed; they are systems that operate as surely as gravity does.

The earth provided a gigantic supply of raw materials that happened to be in the right place at the right time to produce life. And there was time—time for a long slow process to take place—time for random motion to bring a maximum number of molecules in contact with each other.

Some people claim that random chance collision is not likely to present enough contacts for complex molecules to form. This depends on the concentration of the molecules, however. If you shoot a marble through a ring that contains only three other marbles, the chances for collision are few. If you add ten times that number to the ring, the chances of the shooter hitting some marbles are greatly increased, and the coincidental collisions between the marbles ricocheting against their neighbors also are much greater. These physical actions and the chemical processes are functions which take place in nonliving matter. Life was the last function to appear on earth.

NUCLEOPROTEINS

We usually accept the ability to reproduce as the principal difference between the animate and the inanimate. We may wonder where the dividing line will now be drawn between living and nonliving matter because scientists have worked out the way in which supersized molecules reproduce themselves.

We saw earlier that nucleoproteins are formed when amino acids join together to form proteins, and nucleotides join together to form nucleic acids; finally, the nucleic acids join the proteins. So far as we know, nucleoproteins are the only molecules that are able to duplicate themselves.

In view of the complexity of the molecules involved, it is quite probable that variations occurred and differences arose between the old and the new. The probability also exists that identical parts could not be found sometimes, so substitutes were used. None of the molecules were able to make the basic parts themselves; for this, they relied on their environment and the chemical activities within it. To maintain their own structure when parts spontaneously broke down chemically, they had to "feed" on their surroundings.

Slowly but surely the earth changed. More and more nucleoproteins probably formed, once a combination had occurred which was successful in duplicating itself rapidly. Enzymes, the proteins that can act as catalysts without being permanently altered by the chemical reactions they stimulate, were undoubtedly present in the sea. It is quite possible that some nucleoprotein attached an enzyme as a side chain and then kept duplicating this when it duplicated itself. The enzyme would "help" the nucleoprotein by speeding up its activities, and the nucleoprotein would "help" the enzyme by duplicating it.

Which came first, the cell or the nucleoprotein? We know that in the modern system each depends on the other and neither can live for long without the other. Authorities disagree on which came first, and a reasonably good case can be established for either one. Some scholars suggest that the structure of the coacervate cell would have encouraged speeded-up syntheses so that the coacervate would have been the site in which the nucleoprotein developed. Bacteria have nucleoproteins scattered through their cells, but they are not located in a nuclear center as in most other cells.

It is also possible that the coacervate and the nucleoprotein arose separately but simultaneously. The coacervate that incorporated a nucleoprotein (or the nucleoprotein that incorporated a coacervate) had a distinct advantage over either one that was existing separately. Competition challenged all of these macromolecules at the same time. In their own extremely limited environment, the supply of essential materials could have run out long before the total earth supply was exhausted.

The first living things thus may have been the coacervate-nucleoprotein complex that was best able to maintain a steady inflow of materials and energy without destroying itself; one that could grow, could divide, and pass its enzyme system along to its duplicate. Life is more than the ability to divide into equal and identical parts. Maintaining a stable state of continuous building-up and tearing-down, of balancing the gain and the expenditure of energy, is an essential requirement for life.

Viruses are common modern nucleoproteins but they have never had, or have lost, the ability to reproduce. They cannot live in a free state. The host cell furnishes the materials for the virus to make a replica of itself. Apparently, the cell is unable to distinguish between its own DNA and the virus nucleoprotein.

THE CHANGING ATMOSPHERE

At some time in the ancient sea, the food supply of hydrocarbons began to diminish. We do not know exactly what caused this, but the demand began to exceed the supply. Eventually, changing conditions prevented the formation of methane and ammonia in the atmosphere. Perhaps ultraviolet radiation decreased or lightning discharges were less frequent; perhaps other phenomena caused different sorts of chemical reactions to occur so that the composition of the atmosphere was changed.

Authorities differ on whether or not carbon dioxide was present in the original atmosphere. Some hold that carbon combined readily with the little oxygen that was present to form carbon monoxide (CO), and then carbon dioxide (CO_2). Others believe that the available carbon was united with metals to form carbides that were carried into the interior of the earth, and that the available atmospheric oxygen probably united with hydrogen to form water molecules.

We do know that carbon dioxide appeared in the atmosphere. The carbon dioxide might have been formed as the end product in the transformation of the hydrocarbons discussed earlier. Methane (CH_4) can replace each of its four hydrogens with other atoms or groups of atoms, releasing energy by oxidation. The end of the oxidation of sugars, for example, occurs when all of the hydrogens linked to a carbon have been replaced with oxygen, and all the available bonds are tied up. This forms carbon dioxide, which has no more energy to contribute by a transfer of electrons. A similar reaction could have occurred in the ancient sea of molecules; it occurs now inside the living cell when sugars are metabolized to provide energy for the cell.

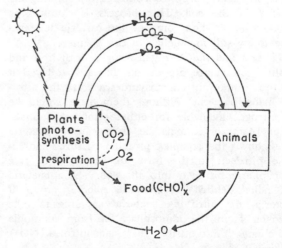

Fig. 11. Production of Carbon Dioxide

While we have greatly simplified the complex chemical actions involved, we can discern a possible source for a new atmosphere composed of carbon dioxide. When the atmosphere changed, the macromolecule cells were forced to seek food.

You have observed a drop of water surround a small bit of matter by flowing around it and then forming into a droplet again after it encircles the particle. In the same fashion, one protein blob could easily engulf a smaller one, then use the engulfed matter to add to its own structure. Some modern single-celled animals live in this way, so perhaps the primitive coacervate cell began to engulf other cells.

If the cells fed on one another, it merely postponed rather than eliminated the need for finding a new food source. Carbon must be obtained to be burned as fuel (oxidized), and nitrogen must be obtained to build the complex protein molecule body for life to continue. Yet hydrocarbons were disappearing from the sea as methane disappeared from the air. The supply of ammonia, the common source of nitrogen, was also diminishing.

If something new had not developed, the adventure of life would have been concluded without a trace. The nucleoproteins and coacervates would have broken down as soon as the supply of new parts gave out. Fortunately, some interesting variations in the large molecules had developed while there was still ample food to allow all kinds of variations to survive.

CHEMOSYNTHESIZERS

Some of the primitive bacteria apparently developed a new way to secure energy by splitting inorganic molecules of iron and sulfur compounds. Some obtained energy from nitrogen by splitting the nitrides in the soil. This process is known as **chemosynthesis,** and some of the bacteria in existence today still metabolize in this manner.

Bacteria and the Nitrogen Supply. Both iron and sulfur bacteria are present on the earth today, but the most important chemosynthesizers are the **nitrogen-fixing bacteria.** Without them there would be no nitrogen available for either plants or animals. When plants or animals die, decay-causing bacteria break down the complex protein structures and release nitrogen in the form of ammonia or free atmospheric nitrogen into the soil. These bacteria are called **denitrifying bacteria.** Another group of bacteria, the **nitrifying bacteria,** are able to take nitrogen from the atmosphere or from ammonia and change it into nitrites (NO_2) and nitrates (NO_3) by oxidation.

The only usable source of nitrogen in the modern world is in the form of nitrates that can be drawn into plants through the roots. The only way animals secure nitrogen is for them to eat the plants or to eat animals that have already eaten plants. No animal or plant is capable of using nitrogen gas directly from the atmosphere.

AUTOTROPHS

The chemosynthetic organisms developed enzymes that enabled them to use chemicals for energy. Other organisms developed an enzyme system of pigments that would "trap" sunlight (visible light) as a source of energy instead of using the ultraviolet radiation that had eventually been greatly reduced. These organisms are known as **autotrophs** because they build their own food instead of getting it from their surroundings.

We call light-trapping organisms **plants** and we can recognize them readily because of their pigments. **Chlorophyll,** the green pigment present in many plants, is a complex molecule built of an entire ring of other rings formed by carbons, nitrogens, and hydrogens, and carries a "tail" of carbons, hydrogens, and oxygen. In the very center of the main ring is a single atom of magnesium. The first "plants" were, of course, not like the present ones, but were similar to the other cells or blobs of protein swimming in the sea.

PHOTOSYNTHESIS

While we are not concerned here with plants as such, it is important to recognize their role in supplying food for all organisms, as well as their production of oxygen. In the primitive world, the differences that we now see between plants and animals had not yet arisen. The simple blobs of protein which developed chlorophyll did, of course, eventually lead to the formation of the plant kingdom.

Chlorophyll is a pigment giving color to leaves, but it also functions as an energy-capturing device. The pigment is arranged in a lipoidal (fatlike) layer, between two layers of protein. Many of these layers, or **grana,** are arranged in stacks, forming a structure called a **chloroplast,** which is surrounded by a membrane inside the cell. The number of chloroplasts varies with the complexity of the plant.

The energy of light excites the electrons in the chlorophyll molecule and, in some way that is not yet well understood, changes the energy received into chemical energy. Carbon dioxide is taken into the chloroplast and hydrated to form sugars, the carbohydrates. In this process water is oxidized by the removal of hydrogen, and the carbon in CO_2

is reduced by receiving the hydrogen to form $(C \cdot H_2O)$:

$$CO_2 + H_2O \xrightarrow{\text{light}} (CH_2O) + O_2$$

According to recent research, the same substance that acts as an electron carrier in animal cells, nicotinamide-adenine-dinucleotide (NAD), may perform this function in plants. The details of this activity in animals will be seen later.

Surprisingly, the complex chlorophyll molecule is quite similar to hemoglobin, the red molecules of the pigment heme combined with the protein globin found in the blood cells of many animals. In animals, hemoglobin carries the oxygen supply to the cells and the blood returns to the lungs with a load of waste carbon dioxide—the chemical that is used by chlorophyll in plants. (Hemoglobin has an iron atom in the center of its ring of rings, instead of the magnesium found in chlorophyll.) See Fig. 12.

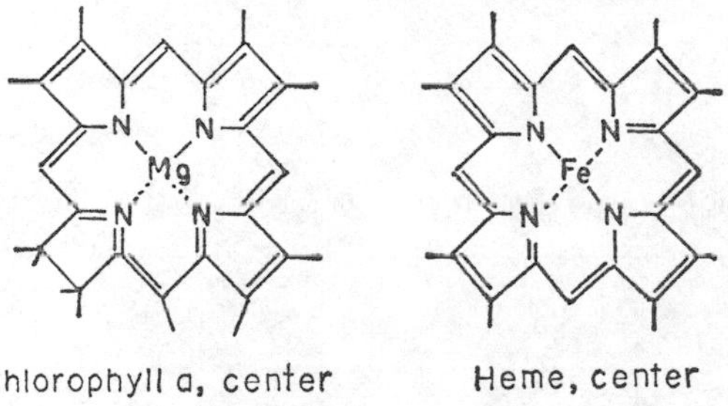

Chlorophyll a, center   Heme, center

Fig. 12

HETEROTROPHS

Organisms that depend on other organisms for food are known as **heterotrophs,** or "other feeders." There are many metabolic pathways for the food gained to take, in providing the organism with energy for maintaining itself. These will be discussed later.

Actually, we must consider the very first living systems as other-feeder heterotrophs because they did not manufacture any food for themselves. They simply gathered whatever spare parts they happened to need from the soup surrounding them and are, therefore, called **primary heterotrophs.**

The heterotrophs that came later are called **secondary heterotrophs.** They feed on other organisms that have managed to manufacture or to gather the essential food. Probably the secondary heterotrophs developed soon after the primary ones and possibly before the autotrophs developed pigments as a way to manufacture their food. Secondary heterotrophs simply transferred their attention to the autotrophs when most of the primary heterotrophs disappeared.

A NEW ATMOSPHERE WITH OXYGEN

One of the end products of photosynthesis is free molecular oxygen (O_2), and the increase in the number of autotrophs began to change the characteristics of the atmosphere again. Not since the first stages of the formation of the earth had any free oxygen been present in the environment. The oxygen released by autotrophs would have reacted with any remaining methane in the atmosphere to form more carbon dioxide which would be used for more photosynthesis. Oxygen would also have reacted with any remaining ammonia to produce free nitrogen and water.

What is the composition of atmospheric air today? It is about 78 per cent nitrogen, 21 per cent oxygen, and .04 per cent carbon dioxide. The rest is composed of small amounts of various gases. Water vapor is present in varying amounts. The relatively large quantity of oxygen that began to accumulate through the activity of the autotrophs would allow the development of organisms which must take oxygen into their structures directly, as animals do when they breathe. Plants breathe oxygen also, but the quantity is small compared to that required by animals.

The use of carbon dioxide by plants is nicely balanced by the production of carbon dioxide as an animal waste product. Conversely, oxygen is given off as a waste product of plant photosynthesis, and is, in turn, used by the animal world. The nitrifying bacteria maintain a supply of nitrogen in usable form which is adequate for plants and animals. No doubt this happy balance came about accidentally, through slow, hesitant steps; the systems that survived show admirably the remarkable balance necessary to maintain a total stability in life as we know it.

TRANSITIONS

The tremendous amount of time that was required for all these changes that took place should be emphasized. If the earth is about five billion years old, it apparently took about half that time for a system that could be termed a "living system" to develop. Small wonder, then, that chemists have trouble trying to duplicate the synthesis of life in the laboratory, even with advanced knowledge of biochemical reactions and catalysts.

It seems logical that the first round of reactions in the newly formed earth was simply the combining of atoms into molecules such as methane, ammonia, and water. The second step must have been the formation of hydrocarbons such as simple sugars, alcohols, and amino acids. During the third major stage the large, complex molecules developed, such as the macromolecule polysaccharides (multiple sugars) and the proteins built from amino-acid peptide chains.

The fourth, and the most important, development to that time was the formation of the nucleoproteins which had the ability to duplicate themselves. Following this, the coacervate colloid cluster formed a cell as the fifth step toward a stable life system.

The need came then to find another source of food because the atmospheric soup was rapidly becoming thinner. With the advent of the living systems that were able to manufacture food and energy for their own needs, we arrive at the beginnings of the life we know today: the chemosynthesizers, the autotrophs, and the secondary heterotrophs are ready to develop the fantastic array of organisms that we see about us in the world.

CHAPTER 3

SIMPLE LIVING ORGANISMS

THE UNICELLULAR WORLD

A whole world of living organisms exist in which the inhabitants are not visible without the aid of the microscope. There is great variety in this world, for it includes the single-celled animals (or protozoa), the single-celled plants (algae, yeasts, and molds), and the bacteria and viruses. Furthermore, while it is easy to distinguish between living and nonliving matter and between plants and animals in the visible world, there are few obvious distinctions at the level of these micro-organisms.

Characteristics of Living Organisms. Generally, living matter is characterized by the ability of the organism to metabolize, to grow, to reproduce, and to respond to stimuli. **Metabolism** is the ability of an organism to take in materials and energy, to incorporate these or convert them to other forms of energy, and to eliminate waste materials. **Growth** may involve an increase in the size or in the complexity of the living organism. **Responsiveness to stimuli** varies in degree with the complexity of the living organism, but is present to some degree in all living things. The **ability to reproduce** is probably the most basic characteristic of any living organism.

Viruses: Living or Nonliving? Viruses (Fig. 13) exemplify the borderline between living and nonliving matter. They are complex nucleoproteins, but appear to be metabolically inert when in a free state. Inside a host cell their activities may cause considerable disruption to the host, producing a variety of diseases such as poliomyelitis, rabies, and viral pneumonia in animals, as well as a number of plant diseases. We cannot accurately define the viruses as "living" despite this activity, for the viral nucleoprotein is not capable of reproducing itself but is dependent on the host cell to perform its reproduction for it. In effect, the virus serves as a false template or pattern for the reproductive mechanism of the host cell; the host cell keeps duplicating this foreign invader every time it duplicates its own genetic template. The virus may be considered parasitic although in all other cases of parasitism the host provides nutrition and protection while the invader is usually extremely well endowed with reproductive apparatus to ensure its own survival.

However, viruses probably originated at about the same time as cells.

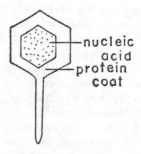

Fig. 13. Virus

We cannot interpret any presently existing virus as being like the first complex protein molecules that progressed from the nonliving to the living, since modern viruses are all dependent upon a host. Nevertheless, the simplicity of these "organisms" may give us some idea of the nature and level of organization of the earliest living matter.

The nucleic acids constitute the bulk of a virus, covered only by a protein coating and tail. This nucleoprotein combination probably formed the basis of the earliest living organisms. In progressing from this level to a truly cellular form, the emphasis was apparently on the organization of the nucleic acids since these self-replicating molecules carry the information that directs the activities of the cell; they are the genes. The simpler viruses contain RNA while the more complex contain DNA.

THE MONERA

In some of the first primitive cells, the nucleoproteins were scattered about or clumped together throughout the cell. This nuclear material was mixed in with the rest of the cell contents. It is thought that the bacteria (Fig. 14) stem from this rather elementary level of existence since they are considered to be the simplest living cells today.

Fig. 14. Bacteria

As we have seen, some of the primitive bacteria were probably able to synthesize food directly from inorganic molecules such as iron, sulfur, and nitrogen in their surroundings. Some of the present-day bacteria retain this ability to chemosynthesize. But most modern bacteria have become dependent upon other organisms to furnish their food. They are **saprophytic,** which means they feed on the bodies of dead plants and animals, causing the breakdown of these structures in the process known as decay.

Another group of organisms of similar level of development are the blue-green algae or Cyanophyta. They contain unique pigments for trapping light energy, phycocyanin and phycoerythrin, as well as chlorophyll scattered throughout the cell. The bacteria and the blue-green algae are sometimes referred to as the **Monera.** They apparently departed on their separate lines of evolution very early, before the unicellular organisms began to organize the nucleoproteins into a structure known as the **nucleus.** The nucleus usually lies in the center of each cell, surrounded by a **nuclear membrane** that separates it from the rest of the cell.

THE PROTISTANS

All of the unicellular organisms that have evolved with a definite nucleus undergo a specialized method of sorting the nuclear material into threadlike chromosomes when the cells divide to reproduce. This is known as the process of **mitosis,** which will be discussed in detail in Chapter 5. Mitosis has never been observed in the Monera.

A large number of variations must have occurred during the formation of the early organisms. Various pigment molecules developed, food was stored in many different forms, and protective and stiffening coatings were developed from various substances. Evidences of these developments can be seen in the modern unicellular organisms.

The division between plants and animals is readily apparent when one distinguishes between a tree and a cow. Plants are usually anchored securely, while animals are mobile. Plants usually are green and make their own food, while animals do not show these characteristics. These traits, which enable us to distinguish easily between plant and animal in the visible world, are not nearly so constant in the world of micro-organisms. Here, both plant and animal may be unicellular and mobile, may contain chlorophyll or other colored bodies (chromoplasts) which perform food synthesis, and may store their food in a variety of ways.

Because of the overlapping of characteristics at this extremely simple level, some authorities have proposed the establishment of a kingdom called the **Protista** from which both plant and animal kingdoms are derived. While this system is not often used formally, it is a valuable concept since it indicates the numerous variations which were present among the primitive organisms. Some of these variations were established later by natural selection in one direction of evolution but not, perhaps, in another.

It is necessary to keep in mind that no organism existing today is the same as any organism which existed at the beginning of life. If we think of the evolving world of living things as a growing bush we can visualize the relationships between organisms. Probably life arose in a variety of combinations, the roots representing the transition between nonliving and living. Various stems survived, developed, then branched out at different levels. Some branches died before the bush reached its maximum height, while some stems grew up with few offshoots and little change. If we look down on the top of a newly clipped shrub we see the ends of many stems, some large and some small, but we will rarely find any branches joined together at that level. An examination of the record of geologic history is like making a cut with the hedge shears at a lower level on our shrub—we may disclose some short branches which stopped growing at a lower level and we may find larger undivided stems from which various twigs branched at higher levels.

Nutrition among the single-celled organisms ranges from clearly plantlike to clearly animal-like. **Holophytic,** or autotrophic, **nutrition** is exactly like photosynthesis in plants, in which the chloroplast structures in the cells utilize the water and carbon dioxide from the atmosphere and the energy from the sun to manufacture carbohydrates. Proteins are built by combining nitrogenous salts which are dissolved in a surrounding medium with the carbohydrates. If complex substances such as amino acids, peptones, and sugars are dissolved in the medium, some protists absorb these directly and are said to be saprozoic. True animal nutrition is **holozoic,** meaning that plants or other animals must be ingested. Although animals may feed on other animals, someplace in the sequence of feeding there is an animal which feeds on plants, so all living organisms are dependent upon the production of carbohydrates by plant photosynthesis.

The current protists are probably different from their ancestors in many ways, but they have not increased in complexity in the same way as other groups have that evolved from them. A microscopic single cell equipped with one or more filaments (flagella) which could be used as whiplike motile organs seems to be the most primitive type of protist. Presumably, these protists could also move about in an amoeboid creeping manner if the flagella were cast off or lost.

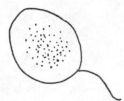

Fig. 15. Primitive Protist

Two characteristics which are generally attributed to plants are the possession of chlorophyll and the presence of cellulose in the cell walls. Among the true green algae, in the phylum Chlorophyta, botanists include both multicellular sessile plants, such as "sea lettuce," and unicellular flagellates and amoeboid organisms. Some of the latter are also claimed as animals by the zoologists and are classified in the phylum Protozoa under the group known as phytoflagellates, or plantlike flagellates.

Volvox. Both botanists and zoologists study the common pond water inhabitant *Volvox* as an example of a unicellular flagellate which forms colonies. True, it has cellulose cell walls and chloroplasts which photosynthesize its food; but its active behavior makes it seem as much like an animal as a plant. Botanists place these volvocids in the order Volvocales, while zoologists catalog them in the

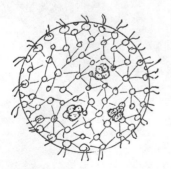

Fig. 16. *Volvox* Colony

family Volvocidae in the order Phytomonadina. The collections of cells which form the colonies vary among the species, with some being little more than loose aggregations of cells. Others are well enough organized to show specializations. The anterior cells of *Volvox* are able to react to light while the posterior cells are the only ones capable of reproduction. See Fig. 16.

Euglena. Botanists sometimes classify the common pond inhabitant *Euglena* (Fig. 17) in a separate phylum, Euglenophyta, since it has chlorophyll. However, it stores a unique animal starch, called paramylum, and it lacks the cellulose wall present in plants. Zoologists classify the euglenoids under the phylum Protozoa as part of the phytoflagellates in the order Euglenoidina. The *Euglena* are obviously closely related to a number of unicellular organisms which are similar in form—some exactly like the *Euglena* except for the lack of chloroplasts. It seems unrealistic to separate these forms widely under such conditions, yet this is commonly done in the traditional concept of plant or animal kingdoms. Thus the usefulness of the protistan concept can be seen.

Fig. 17. *Euglena*

Rather than argue interminably about the problem of where to list or how to list, it is wiser to recognize the limitations of artificial listings and appreciate the shared characteristics shown by the unicellular level of life.

If we consider the unicellular organisms as our first actual living group, we may imagine that in the primitive environment there were many opportunities for success in survival and reproduction. Over many years of reproducing, a number of variations occurred. Some variations were more successful than others in helping to feed and maintain the organisms, so they survived, at least for a time. New variations would continue to occur until finally one group would be committed in the direction of a plantlike existence, another in the direction of an animal-like existence, with many gradations between. Over millions of years the less successful organisms would tend to disappear, making the differences between plant and animal groups more obvious. However, it should be remembered that there are many similarities between higher plants and animals. Many of the chemical syntheses are quite similar, and chlorophyll differs from the oxygen-carrying pigment in the blood of man (hemoglobin) primarily in its possession of a magnesium atom instead of an iron atom at the center of its complex structure.

The structure of the acellular or unicellular organism is considerably more elaborate than that of a single cell which is part of a multicellular organism. Various **organelles** (little organs) have been developed to perform specialized functions in the unicellular animal; these functions would be performed by multicellular **tissues** or **organs** in a multicellular animal. It is difficult to realize sometimes that the tiny protozoan seen zooming about in the field of the microscope produces all that activity within the confines of a single cell.

CHAPTER 4

THE CELL

The cell has been identified for many years as the basic structural unit of living organisms; yet our awareness of this fact depends entirely upon the instruments we have available to view the cell.

The principles of simple lenses were known during the Middle Ages and it is purported that various people had made eyeglasses prior to the seventeenth century. Grinding lenses by hand was a laborious task, requiring considerable patience and ability, and most of the resulting products were undoubtedly of poor quality—more suitable as curiosities than for optical equipment.

Zacharias Janssen, a spectacle maker in Holland, is supposed to have been the first, in 1591, to assemble convex and concave lenses in a tube. When looking through one end of the tube objects were magnified, and when looking through the other they were reduced. Galileo (1564–1642), hearing of this, constructed a "telescope microscope" with which he made observations for his great book on astronomy, *The Heavenly Messenger,* published in 1610. So it has been said that Galileo made the first effective microscope. Numerous lens makers began to construct instruments and make refinements which led to the revelation of a world of life that was invisible before the "magic lenses" were employed.

Van Leeuwenhoek of Holland became famous for his meticulously polished lenses of high magnification and for the observations he made with them. He mounted the lenses in clamps and observed everything he could find with them. Although van Leeuwenhoek was uneducated, his observations were carefully controlled and so well recorded that he was made a member of the Royal Society of London. Over many years, he observed the protozoans that are so prolific in nature; in ponds, in soil, and between his own teeth! He was particularly vehement in attacking the theory of spontaneous generation and he demonstrated many insect life cycles, previously unknown, to prove that life came only from the living.

DEVELOPMENT OF THE CELL THEORY

Robert Hooke, using one of these early microscopes, discovered the boxlike construction units of plants and gave the name "cell" to these structural units in 1665.

The cell theory is generally credited to two men —Theodor Schwann (1816–1882) and Matthias Schleiden (1804–1881). Schleiden was a botany professor at Jena, Germany, for some years. He was impressed by the work of the English botanist Robert Brown on the cells seen in orchids, and developed the concept that all living organisms were composed of cells. He published a paper on this in 1838 in the journal of Johannes Müller, the great anatomist and physiologist in Berlin. It is said that he discussed this idea with Schwann, a student of Müller's. In 1839 Schwann published a more complete cell theory based on his microscopic investigations of both plants and animals. He concluded that the entire plant or animal is composed either of cells or of substances thrown off by cells.

Schleiden recognized the importance of the nucleus, and also discovered the **nucleolus** (*pl.:* nucleoli), the small, granular sphere sometimes seen within the nucleus. Schwann was a shy, easygoing individual and never produced any further research after his university days. Schleiden was tempestuous and unstable, Schwann's opposite. Schleiden's botany text, published in 1842, contained important findings on plant cells and physiology. But after this time he, too, produced little or no research and he was known for the remainder of his life as a strange, warped person. These two men, so very different in temperament, came together under the influence of the great scientist Müller and produced works of major importance. Once away from this influence they failed to produce further work of significance.

To mention only two or three of the pioneers involved in scientific discoveries about the cell is unjust to many others who formed the scientific climate of the times. No theory ever springs up magically or gains adherents immediately. There must be an intellectual atmosphere which permits the new idea to appeal to some influential contemporaries. Usually, there are several people, perhaps independent of each other, who are skirting an idea, on the verge of formulating a new hypothesis, when one of them succeeds in deriving a penetrating analysis or a concrete statement from the nebulous ideas. Each new theory is, in effect, a summation of bits and pieces of knowledge laboriously disclosed by predecessors. The theory could only have developed because the previous work was done.

An example of changes in theory is seen in the one reviewed earlier about the origin of life. We recognize now the fundamental biological principle that life arises from life. But, in earlier times, it was widely believed that "life" arose from various mixtures, usually smelly, which indicates to us that contamination of these messes by various organisms was actually responsible. Redi demonstrated the transformation of maggots to flies, and, many years later, Pasteur demonstrated that contamination by even smaller microscopic organisms caused wines and cheeses to spoil.

Other pioneer microscopists like Marcello Malpighi (1628–1694) and Swammerdam (1637–1680) turned their lenses to anatomical observations, each of them contributing isolated information which would eventually expose the underlying principles of **cytology,** the study of cells.

Rene Dutrochet in 1824 made the statement that plants are composed entirely of cells and of organs derived from cells, and he proposed that this was probably true of animals as well. Rudolf Virchow in 1855 further expanded the cell doctrine by stating that all cells come from cells. This statement was made nearly two hundred years after Hooke had made his observations on plant structure and indicates how slowly observations and ideas evolved. Communication between scientists was sketchy and it was more difficult then to collect observations and evaluate them than it is today.

EARLY CELL OBSERVATIONS

The structure of the cell as it was seen with the aid of the primitive microscope was not complex. But, in many ways, our concept of the cell did not change appreciably through the years of improvement of the light microscope, the instrument in use today. The cell was described as a simple box, with a limiting membrane in animals and a cellulose wall in plants. This figure represented the basic unit of life. Within the box lay a simple, rounded, dark-staining structure—the nucleus. Within the nucleus was a smaller, darker-staining, round nucleolus. In the cell substance, or cytoplasm, outside the nucleus, might be seen a small dot called a **centrosome,** which became active when the cell was ready to reproduce itself.

Some cells showed granules or inclusions in the cytoplasm; sometimes, hollow vesicles called **vacuoles** formed which contained fluid or substances brought into the cell. Occasionally other granules and threadlike structures of unknown function were sighted and named: mitochondria and Golgi bodies, for example. It was argued that these structures did not actually belong to the cell but were foreign matter introduced or artificially produced by the processing undergone by the tissue before it can be viewed on the slides.

The microscope is responsible for a tremendous portion of our present knowledge of biology. Much of the research done with the microscope was in the realm of determining greater anatomical detail, especially among the smaller animals. Another area related to anatomy was the identification of a multitude of species and other taxa for classification. The 1800s were perhaps the most productive years for describing new animals and plants. Collections were made all over the world and classified by the great museums or by the amateur scientists who contributed greatly during this flourishing period.

The knowledge of the cell itself, however, progressed more slowly. Although Robert Brown discovered the nucleus of a plant cell in 1831, its function was not then understood. Investigators later learned that excising the nucleus usually resulted in the death of the cell or the loss of its ability to reproduce.

The function of the **chromosomes,** which were observed as dark-staining strands within the nucleus at certain times, was not understood. It is not known who first observed the strange movements in which these strands engage when the cell divides. Since this process, called mitosis, can easily be seen in cells such as those of onion root tips, probably many investigators saw them and began gradually to speculate on their nature. From the 1870s to the 1920s the realization grew that the chromosomes carried the patterns for cell duplication.

Thomas Hunt Morgan (1866–1945), an American zoologist, and his co-workers discovered the large chromosomes in the fruit fly *Drosophila melanogaster* and began to experiment with them. It was soon shown experimentally that certain loci on the chromosomes controlled certain characteristics. These loci, named **genes,** could be mapped by making many cross-breeding experiments which indicated that the gene was in some way a controlling entity. The small size of the gene placed knowledge of its structure beyond the capabilities of the light microscope, however, so it was not revealed until the electron microscope was developed between 1940 and 1945.

ULTRASTRUCTURE OF CELLS

If the world revealed by the light microscope is impressive, the advent of the electron microscope has opened another world of amazing complexity which lies beyond the magnifying capacities of the earlier instrument. Structures of almost incomprehensible smallness, yet fantastic intricacy, can be observed. The visible has reached the level of the chemical, and the knowledge of the biologist overlaps that which the chemist has learned through experiment and the physicist has learned by measurement.

Electron Microscope. Resolution (or resolving power) is the ability to distinguish the parts of an object. The lower limits of resolution are the smallest size at which two adjacent points can be brought into focus as separate objects. The resolving power of the human eye ranges from .025 to .1 millimeter. The best light microscope, which focuses by passing light through focusing lenses, can resolve down to one half the wave length of light, .275 microns or 2750 Å. ($\mathring{A}$ is the symbol for Angstrom unit; 10,000 Å=1 micron=.001 millimeter.) The electron microscope is now able to resolve down to about 10 Å, although its theoretical possibilities allow a magnification down to 0.05 Å; the hydrogen atom is 1.06 Å. So the electron microscope, which sends a stream of electrons through electromagnetic focusing "lenses," is at least 10,000 times more powerful than the human eye.

The phase microscope, although a light microscope, employs a system of breaking up the light beam so that a portion of it detours and arrives at the object being examined somewhat delayed and at an angle. Living material can be observed by the contrast in light imposed, whereas in the conventional light microscope tissues usually must be treated with selective stains to provide contrast. The light microscope is about 500 times more powerful than the human eye.

The Shape and Size of Cells. While some cells may give the impression of being box-shaped or round, most of them are actually many-sided (polyhedral) because of the compression exerted by adjacent cells in multicellular structures. This provides both an economy of space and at the same time a maximum of surface for contact between cells in order that various substances may be exchanged. The vast majority of cells fall within the size range of 10 to 100 microns (μ), which probably indicates the practical functional limits for nourishing the cell.

The *surface* of a sphere increases in proportion to the radius squared, while the *volume* increases in proportion to the radius cubed. Consequently, in a growing cell, the volume would soon exceed the ability of the cell membrane to furnish diffusion surface, the probable way by which the cell obtains most nutriments and rids itself of wastes. The surface of the cell could also be increased by changing the shape of the cell by elongation or branching. A variety of cell shapes exist, but changes in shape are probably primarily associated with the functional duties of the cell rather than with diffusion surface.

THE CELL MEMBRANE

The plasma membrane that forms the boundary of the cell is extremely thin; yet, as electron micrographs show, it is multilayered and, in some cases, folded into **microvilli** (small, finger-like projections) to increase the surface. The properties of the cell membrane are exceedingly important because they must protect the functioning of the cell, and thus maintain the stability, or **homeostasis,** of the organism. Stability is not a passive state in the organism—continuous activity is necessary to maintain the stable state, balancing intake and output.

The cell membrane has been analyzed as being a double layer of phospholipids coated on both outside and inside by proteins. In electron micrographs, this appears as three parts: a dark line, then a light space (or layer), then a dark line. "Pores," or breaks, occur in the protein surface; yet the size of the pores alone does not determine the materials or molecules which enter the cell through them. The nature of proteins and phospholipids also influences the intake.

Food for Cells. Proteins are composed of carbon, hydrogen, oxygen, and nitrogen, and, usually, phosphorus and sulfur, linked into simple units called amino acids. Each amino acid has an amino group ($-NH_2$) and a carboxyl group ($-COOH$), plus side chains which form the characteristics of the individual amino acid. The $-NH_2$ group has a positive charge and acts as a base which can combine with acids, while the $-COOH$ group has a negative charge and acts as an acid which can combine with a base. When amino acids link together, the $-COOH$ of the first acid joins with the $-NH_2$ group of the next acid. The $-COOH$ contributes an $-OH$ and the

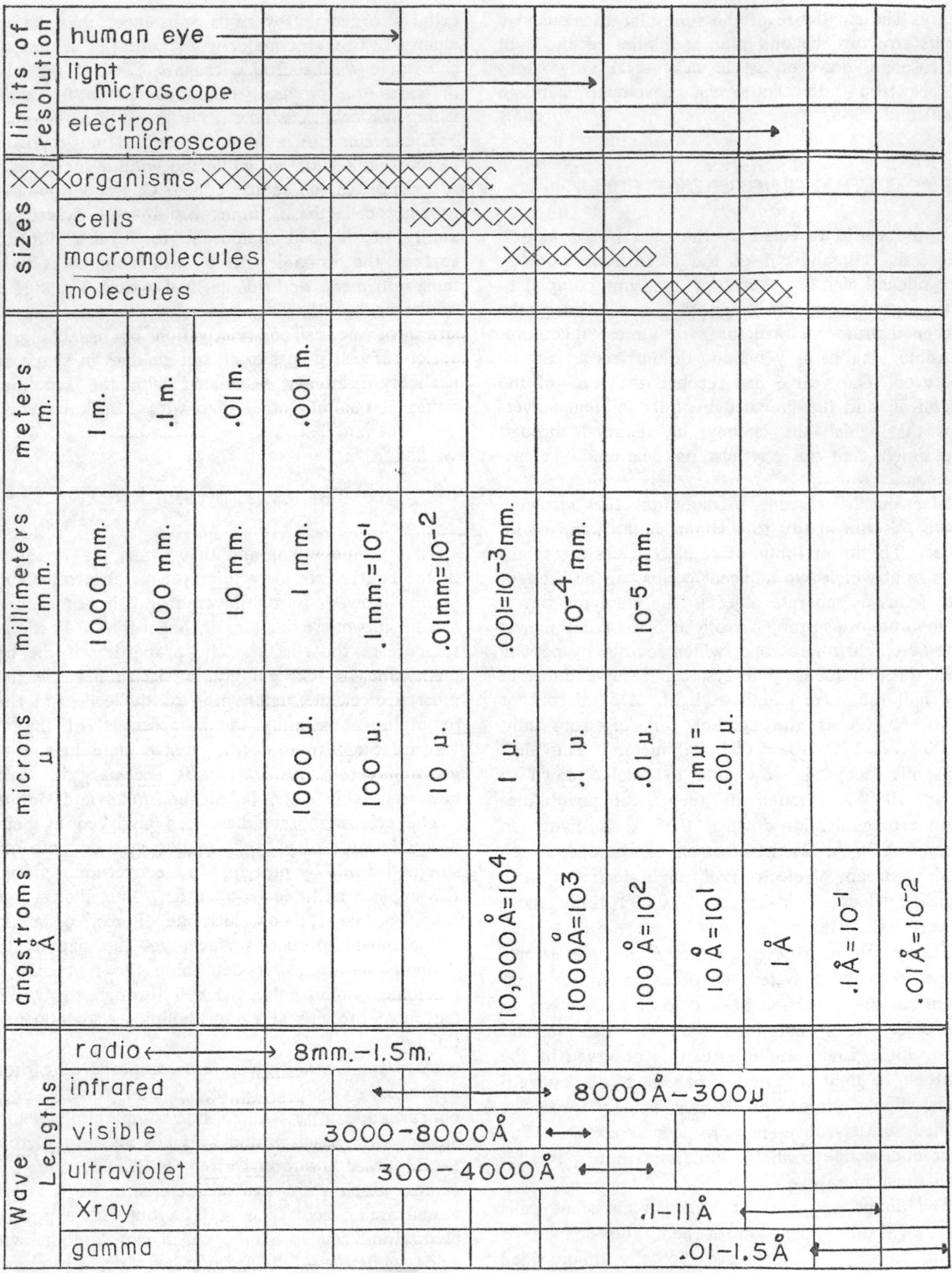

limits of resolution

- human eye ⟶
- light microscope ⟶
- electron microscope ⟶

sizes

- XXX organisms XXXXXXXXXXXXX
- cells XXXXXX
- macromolecules XXXXXXX
- molecules XXXXXXX

meters m.

1 m.   .1 m.   .01 m.   .001 m.

millimeters mm.

1000 mm.   100 mm.   10 mm.   1 mm. $.1 mm.=10^{-1}$ $.01 mm.=10^{-2}$ $.001=10^{-3} mm.$ $10^{-4} mm.$ $10^{-5} mm.$

microns μ

1000 μ.   100 μ.   10 μ.   1 μ.   .1 μ.   .01 μ.   1 mμ = .001 μ.

angstroms Å

$10,000 Å=10^4$ $1000 Å=10^3$ $100 Å=10^2$ $10 Å=10^1$ 1 Å $.1 Å=10^{-1}$ $.01 Å=10^{-2}$

Wave Lengths

- radio ⟷ 8mm.–1.5m.
- infrared ⟷ 8000Å–300μ
- visible 3000–8000Å ⟷
- ultraviolet 300–4000Å ⟷
- Xray .1–11Å ⟷
- gamma .01–1.5Å ⟷

Fig. 18. The Electromagnetic Spectrum

–NH$_2$ contributes an –H to form a molecule of water which is split off from the amino acids. This leaves a –CO bound to the next –NH in what is called a peptide linkage or an anhydro bond, as we saw in Chapter 2. The importance of the protein molecular structure in the cell membrane is that the free ends of protein molecules are **amphoteric;** that is, they have a positive charge at one end and a negative charge at the other, and can act either as an acid or a base. Some substances would be repelled by these charges so they would be kept out of the cells.

The phospholipids are complex organic compounds which tend to dissolve in fat solvents, such as ether, benzene, or chloroform, but not in water. In forming chains, the long phospholipid molecules are thought to orient themselves somewhat like pickets in a fence so that a polar group which will unite with water is at one end and the other end of the lipid chain will unite with air or oil. Lipids of this sort readily form surface films, or may form a double layer with the lipid chains in the center and the polar group facing out. The polar end may have an –Na+ sodium ion which will be attached to the active surface of proteins to form lipoprotein cell membranes.

The lipoprotein membranes are not restricted to the outside of the cell. It is now thought that the cell membrane folds into the interior of the cell,

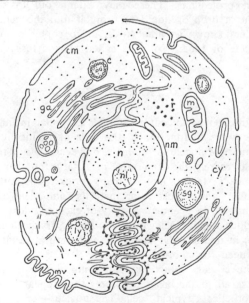

Fig. 19. Schematic Structure of Generalized Cell
c–centriole; *cm*–cell membrane; *cy*–cytoplasm;
er–endoplasmic reticulum; *ga*–Golgi apparatus;
l–lipid; *ly*–lysosome; *m*–mitochondrion;
mv–microvillus; *n*–nucleus; *nl*–nucleolus;
nm–nuclear membrane; *pv*–pinocytotic vesicle;
r–ribosomes; *sg*–secretory granule

forming a network of canals, called the **endoplasmic reticulum,** and possibly continuing inward to form a double membrane around the nucleus.

The Nuclear Membrane. The membrane surrounding the nucleus differs somewhat in appearance from the cell membrane, probably because of differences in the activities going on within the nucleus compared with those carried on in the protoplasm (cytoplasm) that surrounds the nucleus and forms the bulk of the cell. The membrane seems to have pores in it, and it is known that materials can pass through it in either direction. Perpendicular sections of the membrane show thin areas where the two dark layers seem to merge into a single dark thin layer. Tangential sections indicate that these pores may be either open, or closed with a knoblike projection or diaphragm. This may be a selective activity, but how and why the pores open and close is not understood.

The nucleus, which in most cells is visible with a light microscope, is usually the largest of the organelles in the cell. Although it has been recognized for over a hundred years, its function as the center of the reproductive apparatus of the cell was not at first observed. Its principal constituent has been found to be deoxyribonucleic acid (DNA), which is discussed elsewhere.

The nucleus also contains a dark-staining structure, the **nucleolus,** which disappears just before the cell divides to reproduce and reappears afterward. It seems to function in synthesizing ribonucleic acids.

Cytoplasmic Structures. The remaining protoplasm of the cell, the **cytoplasm,** appears in the light microscope as a relatively clear substance with small particles suspended in it. One structure which is present in animals and in lower plants is the **centrosome**—a structure containing two **centrioles,** small bodies which seem to be the kinetic or activity centers. When the nucleus is ready to duplicate, the centrioles separate and, in some cases, seem to exert a definite pull on the nuclear material, drawing it into the pattern for division. The centriole is usually composed of filaments grouped in a circle, nine groups of three filaments each, very much like the structure of the protistan flagellum. The two centrioles are contained in the centrosome with the filaments of one centriole at right angles to those of the other, until they are released and go to opposite ends of the cell during cell division. After cell division the centriole duplicates itself and the centrosome covering forms again.

Mitochondria are small bodies which appear as rods or granules in the cytoplasm and contain numerous enzymes. For many years these and other inclusions in the cells were said to be foreign bodies introduced by the killing, staining, and sectioning techniques used to prepare tissue for microscopic examination. The electron microscope shows mitochondria to be normal cell components. They are sausage- or rod-shaped structures which have a double membrane and internal villi, or folds, called cristae. Each mitochondrion is an energy factory for the cell. Hundreds of mitochondria may occur in a single cell and they are usually distributed around the centers of high metabolic rate in the cell.

By crushing cells and spinning the contents in an ultracentrifuge, biochemists have succeeded in separating out mitochondria and then making cultures of them to determine their activities. They are composed of proteins, phospholipids, and ribonucleic acids. Lying in or on the cristae are complex groups of enzymes which stimulate the metabolism of carbohydrates, fatty acids, and amino acids into carbon dioxide and water with a release of energy by the flow of electrons. This is known as cellular respiration.

The Golgi apparatus appears in electron micrographs as a stack of smooth, flattened vesicles or membranes, sometimes widened into vacuoles. They were first discovered by the cytologist Camillo Golgi in 1898, and were seen as dark rods when stained with silver salts. Cytologists now think that the Golgi apparatus is a part of, or is connected to, the endoplasmic reticulum but lacks the ribosomes, small dark granules of ribonucleoprotein which dot the endoplasmic reticulum membranes. Ribosomes also cluster in the cell away from the reticulum and form the centers of protein synthesis. Enzymes are formed which carry out, in conjunction with ribonucleic acid, the characteristic activities of the cell other than the genetic functions of the nucleus. The endoplasmic reticulum and Golgi bodies are consequently said to have secretory functions. Secretory granules called zymogen granules can sometimes be seen in the Golgi complex. All of these reticulate structures are sometimes called microsomes since they are lighter in weight than the mitochondria and can be separated from them by centrifuging the crushed cell material.

Lysosomes are bodies which vary in size; they do not show the same internal structure of mitochondria of similar size. Biochemical analysis indicates that lysosomes contain hydrolytic enzymes; that is, substances which are capable of breaking down materials, whether from other cells or from within the cell.

The cytoplasm of the cell exhibits movement, sometimes called cyclosis, almost continually. But though the cell structures seem to be loose and wandering, the reticulum apparently controls or connects all of them and preserves an orderly structural unity.

Fibrils that can sometimes be seen in the cytoplasm of certain cells are composed of protein, and may perform contractile or neurocontractile functions. (We cannot properly call these muscles or nerve-muscles, since such structures are multicellular.) Other structures may be present in specialized cells according to the function of the particular cell, as we shall see later.

Other substances, not integral parts of the cell, may also be present. These inclusions are called ergastic substances, and may include granules of starch or protein, or globules of lipids (fatty material).

Cells in the multicellular animal link together first to form tissues. Tissues may then be joined to form organs in all but the lowest levels of multicellular animals. Contacts between cells take a variety of forms. While the cell membrane may appear to be smooth and straight under the light microscope, it may actually be folded into finger-like projections called microvilli. The microvilli greatly increase the surface of the cell membrane, exposing more area for exchange of materials between cells; at the same time they anchor the cells by interdigitation, integrating the fingers from two cells. Other cells attach to one another by bulges, called desmosomes, which look like the rounded projections on jigsaw puzzles, or by terminal bars.

It is one of the properties of cells that they can form projections. These may extend out to surround a drop of liquid to bring it into the cell, a process known as pinocytosis, or to engulf a solid particle of food, a process called phagocytosis.

FUNCTIONS OF CELLS

Function in unicellular animals is similar in many basic ways to function in the cells of multicellular animals. Multicellular animals, however, do not represent merely the sum total of a large number of tiny cells stacked like so many bricks. The multicellular animal must co-ordinate its component cells so that the independent activity of each cell does not produce chaos. Delegation of duties, or specialization, in multicellular organisms is a natural result of the need for co-ordination.

Unicellular Functions. The unicellular animal shows all of the basic structures that appear in the hypothetical "typical" cell. The nucleus—or nuclei, if more than one is present—is the management center for the organism and contains its genetic material. Cellular respiration is carried out by the mitochondria, protein is synthesized by ribosomes, and the Golgi apparatus produces secretory granules. Some vacuoles form to bring water into the cell or to eliminate it, while others secure and store food substances.

The unicellular animal must perform all of the intracellular metabolic functions that are necessary to a single cell, and it must also perform the functions we commonly associate with co-ordinated activity in more complex animals: locomotion, feeding, defense, and reproduction. These functions are carried on by specializations of the structures which exist in single cells.

The protein fibrils may be arranged so that they act as supports in the microvilli. Lengthened, the microvilli may possibly have become the filaments such as short cilia or longer flagella that are modified for locomotion. Cytologists have demonstrated that in cross-section the microvilli show dark granules or strands. The structure of flagella or cilia, and also of the centrioles previously described, are much alike. For instance, nine bundles of three fibers each (some scientists say two fibers each and part of a third) form a circle. The center may or may not have two single fibers in it. Additional fibrils form a connecting network beneath the surface in the unicellular animals which have several cilia, while a strong supporting bundle forms in those which have few or a single long flagellum. Apparently, the fibers are able to act as neuromuscular co-ordinators of these organelles.

Movement due to the flowing of the protoplasm is commonly known as **amoeboid motion** (after the unicellular amoeba). Shapeless unicellular organisms extend temporary projections in the direction of future motion, then the rest of the "body" flows into, and merges with, these projections, which are sometimes called **pseudopodia**, "false feet." This motion is apparently based upon the characteristic ability of proteins to change from a solid to a liquid and back, as gelatin may be in either sol or gel states. Gelatin depends on temperature change to determine its state, but amoeboid cells can change their form rapidly by means which are, so far, unknown.

The protective function in some of the unicellular animals is accomplished by secretions from the cell membrane which form a coating, or **cuticle,** around the cell. Some of the coatings become rather elaborate, with spines, meshes, or plates composed of polysaccharides, calcium carbonate, or silica. These may also serve as **flotation devices,** structures that help the animal float. Naturally, this aids in dispersal of animals with otherwise limited mobility. Plants also exhibit flotation devices.

Unicellular animals are more often multinucleate than are the individual cells of multicellular animals. Perhaps this is because of the many activities which must be directed within that small bit of protoplasm. Some protozoans have two kinds of nuclei: one to control maintenance activity; the other to perform reproductive activities.

Sensory perception is shown even at the lowest level of animal life. A number of the unicellular animals contain pigment spots which are light-sensitive, enabling the animal to orient itself toward or away from a light source. This implies some sort of neuromuscular response to the stimulus, either attraction or repulsion, and the ability to carry out correlated movements. We cannot easily envision any single cell of a multicellular animal performing all of these diversified activities.

Multicellular Functions. Specialization is a way of life, even in the single-celled, or acellular, animals. We know that the cell as a unit contains many smaller structures; therefore, certain areas of the cell must perform certain tasks.

From having specific areas perform specific tasks to having certain cells perform certain tasks is a logical step but a longer one than it first appears because of the vital requirement of **co-ordination—** a function usually carried on by some type of neuromuscular structure. Co-ordination is not easily achieved; perhaps, then, there is some intermediate stage in multicellular life.

Colonialism may be the step which bridges unicellular and multicellular existence. The *Volvox,* discussed earlier, and its relatives provide us with an example of a simple colonialism that seems to progress in the direction of co-ordination between cells. While we must keep in mind that *Volvox* is plantlike in its nutrition, we may examine it assuming that similar ancestors may have existed for an animal line protistan.

The colonial forms show specialization by having **polarity;** that is, an animal pole, or end, which seems to perform the directing and sensory activities, and a vegetal pole, which concentrates on feeding or reproductive activities. The flagella of all the cells are co-ordinated to beat together so that the organism travels, after a fashion. Protoplasmic connections may be seen between the cells in some of the colonial phytoflagellates. They are probably present in all of them, but may be too small to have been observed as yet.

A colony of one-celled animals is probably a **mutualism,** or banding-together, because it is beneficial to the cells involved to be grouped for increased mass, better locomotion, or reproduction. In *Volvox,* the offspring (called **daughter cells)** are extruded into the center of the hollow-ball colony and, thus, are protected.

These examples of colonialism concern unicellular organisms, but colonialism is not limited to them —it is found on all levels of complexity in animal groups. There are great differences in the degree of colonialism; the volvocids are physically linked to each other, but the members of an ant colony obviously are not.

Another reason for the mutual linking of cells may be that although cell division occurs, the resultant cells do not separate completely. Due to the surface-to-volume ratio of a sphere, the practical size to which a single cell can grow is limited. The unicellular organism obtains most of its food and gets rid of its waste products by diffusion through its surface. If the interior bulk becomes too large to have a sufficient diffusion surface, the organism is at a disadvantage. If the cell divides and forms two spheres instead of one large one, the surface is increased and the interior of each cell is closer to its surface membrane.

Cell Linkage in the Sponge. The simplest multicellular animals are usually considered to be the sponges, which will be examined in more detail later. Concerning the advance of unicellular organisms to the multicellular level, the sponges show cells linked in perhaps the most elementary form. The outermost layer of the sponge consists of cells shaped like polygons. This **epidermis** comes closer to being a tissue than any of the other cells, but all the cells of the sponge exhibit independence of one another in performing the basic functions.

Beneath the epidermal cells lies a gelatinous matrix in which wandering cells move around in amoeboid fashion, performing several functions, such as secreting needle-like supportive spicules, transporting or digesting food, and forming germ cells. The inner layer of cells resembles certain flagellate protozoans, but they are not co-ordinated with each other in beating the flagella, although they do produce a current in the water flowing through the sponge. These cells also feed, but within themselves individually; they do not secrete any digestive fluid into a cavity, as a gastrodermal digestive tissue would.

In short, then, sponges are multicellular, but on an extremely elementary level. This structure was demonstrated years ago by forcing living sponges through a silk mesh to separate the cells. The cells continued to live and managed to collect into clusters, forming new sponges. No neural connections have ever been positively located in sponges. Some have muscular fibers in the epidermal cells which enable each cell to contract individually, thus decreasing the total size of the animal. Without intercommunication, it is obvious that the sponges would be very limited in activity; they are able to move only when they are small embryos and become sessile, or attached, as adults. Anarchy among cells could hardly produce an organized, co-ordinated movement. Sponge cell independence restricts the size and complexity of the animal.

SPECIALIZATION OF TISSUE

Sooner or later, cells began to co-operate in their functions, so that true **tissues** developed. Tissues are formed when similar cells group together to perform a specific function. There may be more than one kind of cell involved in the tissue, and the tissue may perform more than one function, but the principle involved is teamwork between cells.

Each multicellular animal is composed of tissue or tissues that vary according to its type. In the simpler groups the tissues are fewer. Sometimes this means that more functions must be carried on by one tissue. This is really in keeping with the capabilities of single-celled animals, however—simplicity of structure means greater generalization. In short, the fewer cells there are, the more versatile each cell must be in order to perform all the required functions. Conversely, the more cells, and the more tissues present, the more specialized each can be because there are other cells and tissues to do the other jobs.

Organs and Organ Systems. The tendency to specialize appeared early in the living world, and the trend has never been reversed. Animals which were composed of tissues began to develop complexes of these tissues grouped so as to perform a given function better. These complexes are called **organs.**

The first organ to appear among the so-called lower animals was reproductive in function. Survival is always keyed to the animal or plant most efficient in reproducing itself. If the organ structure performed a more efficient reproduction function, the animal clearly had an advantage over its contemporaries.

The development of organs increases with the level of organization of the animal group. Finally, organs began to be co-ordinated in function either by neural or secretory means, so animals developed organ systems.

The organ groupings perform the major functions of the animal body and are usually divided into these systems: reproductive, digestive, respiratory, excretory, neural, muscular, skeletal, and circulatory. Other systems may occur such as secretory or endocrine systems; some systems may be combined into one, such as neuromuscular.

Each group of animals carries on the various systemic functions of life in its own way. It is impossible to give a single explanation that would in any way be representative of the function of a given system in all animals. For this reason, the systems will be discussed as each group of animals is discussed.

HISTOLOGICAL SPECIALIZATION

The statement that cells unite into tissues and tissues into organs is a generalization, useful in pointing out the progressive organization of body structure. The tissues, no matter what their function, are built from relatively few types of cells. These cell types often occur in several different tissues or organs.

Functions of Cells. Cell activities may be divided into three main categories. The first is the **vegetative** function, in which cell maintenance is carried on by respiration, nutrition, and excretion, to provide the stable state called **homeostasis.** The second function is **reproduction and growth.** This provides for the increase in cell size and the mechanism for dividing or duplicating the cell. The third function is that of a **specialty,** wherein the cell provides some service to the whole organism over and above its own maintenance and reproduction. The specialized muscle cells provide contraction; the nerve cells furnish transmission of sensory and motor messages; the other cells provide secretions, and so on.

Because of the tremendous increase in research and knowledge of the cell in recent years, **cytology,** the study of cells, has become a very active field. **Histology,** the study of cells in tissues, has taken on increased importance as functional knowledge is added to the knowledge of form or morphology.

Background of Histology. In France, M. F. X. Bichat in 1801 described the differences in protoplasmic structures of several bodies and parts and classified the tissues into twenty-one kinds. He referred to each one as a *tissu,* an old term for a richly woven cloth. At that time, the cell doctrine had not been stated, but a few investigators had referred to the boxlike units in plants and animals. Bichat did not use a microscope, so he did not realize the cellular nature of the tissues. A brilliant surgeon, teacher, and author, he died in 1802 at the untimely age of thirty-one.

In 1844 a British anatomist, Richard Owen, initiated the term "histology" for the study of the minute structure of animal and vegetable tissues, or microscopic anatomy. The prefix is derived from the Greek word *histos,* meaning "web," which is probably also the root for the French *tissu.*

Schwann's work in 1839 elaborated on the comparative nature of plant and animal tissues. He developed a classification of tissue that is essentially like modern classifications except that he combined muscle and nerve tissues.

Albert von Kölliker, a Swiss physiologist, wrote the first text on histology in 1852 which established histology as a separate scientific discipline. He made extensive studies of the tissues of several animals on a cellular basis.

Techniques of Histological Study. In order to study tissues in any great detail, techniques had to be developed whereby thin sections could be examined under the microscope. Wilhelm His, a professor of anatomy at Leipzig, invented the machine called a **microtome** in 1870. This machine could cut hardened tissue into very thin slices which then would be placed on glass slides and light directed through them under the microscope.

Soon after their development in Europe, microtomes were developed and built in the United States. Charles Minot invented the rotary type in 1886. By the turn of a wheel, a continuous ribbon of slices could be cut for mounting. There are still some of the old Minot microtomes in use in colleges today.

Naturally a number of special methods had to be found to prepare tissue for the microtome. Freshly killed specimens are hardened, or **fixed,** by placing them immediately in chemical solutions; a 4 per cent formaldehyde solution is most commonly used. This also prevents bacterial breakdown of the tissue.

Tissue normally contains a very large percentage of water, which must be removed so that the structure may be filled with some stiffening substance. A series of dilutions of alcohol from weak to strong are applied to the tissue to remove the water. Then the tissue is placed in a melted paraffin bath and allowed to stand until it is thoroughly impregnated. While still warm, the tissue and paraffin are poured into small molds, and are then cooled. The blocks are trimmed so that they can be sliced by the microtome. In the early days a number of stiffening substances were tried, such as whale spermaceti, beeswax, tallow, and cocoa butter. Minot introduced the use of paraffin in the United States.

After the paraffin sections have been mounted on slides by warming, they must be **stained** so that they will show detail. But stains cannot penetrate the paraffin, so chemicals are applied to dissolve it. Since most stains are in aqueous solutions, the tissue is rehydrated in a series of alcohols, strong to weak, and then placed in the stain.

A large number of stains suitable for showing certain structures and details have been developed. Most of these were found by trial and error: pour some on and see what happens. Because of this, the way in which the dyes "take" to some cells and not to others is often not understood.

Most stains are said to be **acid** or **basic,** and the tissues are described as "acidophilic" or "basophilic," depending on whether they are attracted to (*Gr.: philos,* "loving") acid or basic stains and are therefore colored by them. Actually, the stains are neutral salts, but they contain both a positively charged and a negatively charged group in the structure, much as the amino acids do. The acid group is the negative or acid-staining area, and the basic group is the positive or basic-staining area. There are, of course, other staining groups than the acid-basic kinds.

Routinely, histologic sections are stained with both an acid and a basic stain to bring out differences. Hematoxylin and eosin are most commonly used for this. Hematoxylin is basic and stains basophilic tissue blue or purple. Eosin is acid and stains acidophilic tissue pink or red.

After sections are stained, they must again be dehydrated to remove the aqueous stain solution so that the slides can be made permanent. They are passed back through the series of weak to strong alcohol. Then a clearing solution is applied which makes the tissue translucent, and a thin coat of mounting medium is applied to the surface. An optically ground thin sheet of glass, the **cover slip,** is then placed over the entire group of mounted sections, and the slide is allowed to dry. Resin from Canadian balsam trees was the most popular mounting medium for years, but synthetic media have almost completely replaced it.

Tissues being prepared for microscopy are subjected to rather rough treatment; cells can be distorted or foreign particles (artifacts) introduced. Refined techniques, such as vacuum-freeze drying and freeze sectioning, are sometimes used to avoid trauma to delicate tissues.

Structure of Tissues. The basic tissues are composed of cells, and of intercellular material. The term "intercellular material" encompasses a number of different things, such as the **ground substance** in which the cells are suspended or by which they are held together. Depending upon the type of tissue, the ground substance may be fluid, semisolid, or solid. The cells of a tissue may vary widely in morphology and function. Generally speaking, we refer to the functional cells of a tissue as the **parenchyma** and to the supporting elements of the tissue as the **stroma.**

There are numerous clear-cut examples of this. For instance, in the salivary glands the parenchyma consists of glandular epithelial cells which produce the saliva. A framework of fibers forms a scaffolding that supports the cells and a fibrous envelope covers the gland; together, these constitute the stroma.

Basic Tissues. In animals, especially in humans where the tissues have been the subject of intensive study because of their significance in medicine, we usually list five groups of basic tissues. They are: the **epithelial** tissues, the **connective** tissues, the **blood** (considered by some authorities as a connective tissue), the **muscular** tissues, and the **nervous** tissues. This list demonstrates that a tissue is regarded as a unit of similar cells which perform a common function.

Epithelial tissues furnish the limiting membranes and protective linings or coverings of the organs, or they may form secretions as in the glands. The connective tissues are quite versatile, holding the body parts together and supporting and furnishing much of the bulk of the body. Muscle tissue provides motility for the body and responds to the transmissions of the nervous tissues that direct muscle contraction.

Among different animal groups, the exact tissues performing a function may differ. Since most of the histological studies have been concerned with **vertebrates** (animals with backbones, such as man) the divisions of the elementary types of tissues may differ from the divisions found in some of the invertebrate groups. Furthermore, a given type of tissue is not always of the same origin developmentally. Epithelium may be developed by more than one of the germ layers that are laid down as an organism is formed. These germ layers will be discussed in detail in the embryological section, but we should note here that "epithelial" describes the nature of a tissue, not its origin.

EPITHELIAL TISSUES

Epithelial tissues are usually composed of many cells that are fairly regular in shape and closely packed so that little cementing substance lies between them. They may be arranged in sheets to serve as linings of cavities and coverings of structures; in these cases, they have a protective function. Or the tissues may be arranged in cords, follicles, or tubules, and perform secretory functions. Sometimes certain of the cells in the sheets will also manufacture secretions, so the dividing line is not always well defined.

Epithelia in sheets may be further divided on the basis of cell shape and location. A **basement membrane,** formed either by the epithelial cells or connective tissue beneath it, underlies the sheet. If all the epithelial cells are in contact with the basement layer, the tissue is called a **simple epithelium.** If the cells occur in layers, so that only one layer of cells is in contact with the basement layer, the tissue becomes a **stratified epithelium.**

Simple Epithelium Tissue. Simple epithelium may be squamous, cuboidal, or columnar, depending on cell shapes. **Squamous epithelium** (Fig. 20) is thin and flattened, with the individual cell boundaries wavy and obscure. The nucleus is in an off-center bulge. It forms linings where strength is not required but where diffusion through thin membrane is important.

Fig. 20. Simple Squamous Epithelium

Cuboidal epithelium cells (Fig. 21) are more like cubes or blocks than the thin, flat squamous cells. They seem nearly hexagonal when viewed from the surface, and appear as squares in sections because their length and width are about the same. The cell boundaries are often clear, with the nucleus well centered in the cell. If the cuboidal epithelium is formed into a circle, to line a tubule, for example, the cells are pyramid-shaped. Some cuboidal cells show granular cytoplasm; this indicates that they are forming secretions.

Fig. 21. Simple Cuboidal Epithelium

In **columnar epithelium** cells (Fig. 22), the surface shapes appear to be about the same as cuboidal cells, but they are taller than they are wide. The free surface may be modified to form a cuticle which bears hairlike processes that are in contact with nerve endings, forming a sensory epithelium. Columnar epithelium also forms the chief secretory tissues of the body, with various modifications determining the type of secretion.

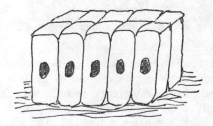

Fig. 22. Simple Columnar Epithelium

Sometimes the columnar cells become so crowded that some of the cells appear to have almost lost contact with the basement membranes, and the cell shapes are distorted. This is called **pseudostratified epithelium** (Fig. 23); its cells frequently have cilia on the free surfaces.

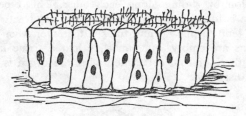

Fig. 23. Pseudostratified Columnar Epithelium

The difference between cuboidal and columnar epithelium is primarily in cell height; therefore the dividing line between cuboidal and low columnar is arbitrary. Remember that the biological organism does not label the categories; it is the people who interpret them who rank the differences in an effort to describe and classify. Intergradations often exist between our artificial categories.

Stratified Epithelium Tissue. Stratified epithelia, the multilayered epithelia, are of several kinds, as the simple epithelia were. The stratified epithelia all have a layer of modified cuboidal cells next to the basement membrane, and above this one or more layers of polygonal cells. The free surface layer of cells is variable—it is this variation that gives us the subdivisions of stratified epithelia: stratified columnar, transitional, and stratified squamous.

Stratified columnar epithelium (Fig. 24) differs from the pseudostratified simple epithelium in that the columnar cells of the stratified epithelium actually have become separated from the basement membrane. A layer of small rounded cells now lies along the membrane and the columnar cells are above the rounded cells. Although this type is rare, it does occur; it lines part of the nasal passages, for example.

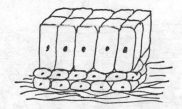

Fig. 24. Stratified Columnar Epithelium

Transitional epithelium (Fig. 25) is composed of several layers. As in stratified columnar epithelium, the cells lying along the basement membrane are small and rounded. Above these are several layers of polygonal cells, the uppermost of which bulge into an inverted pear shape. The surface layer consists of larger dome-shaped cells that cover several polygonal cells. These cells possess an ability to change position by sliding along each other. This type of tissue is found in the lining of the urinary tract. The cells tend to pile up and thicken the layer when the organs are empty, but can slide into a tissue layer only three or four cells thick when the organ is full and distended.

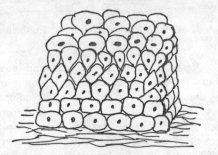

Fig. 25. Transitional Epithelium

Stratified squamous epithelium (Fig. 26) is perhaps the best-known epithelium; it forms the protective skin surface of our bodies. The thickness of the layer and the structures that the outer surfaces develop vary according to the animal and the body region. The basal cells are the usual small, rounded cells, and, with many layers of polygonal cells above, they form the layer called the **stratum germinativum** in man. These cells appear by the light microscope to be connected by many intercellular bridges, and have been called "prickle" cells. Actually, electron microscope studies show that the "bridges" do not connect the cytoplasm but are desmosomes that interlock. As the layers approach the surface, the polygonal cells become more and more compressed until they resemble simple squamous epithelium. They differ in that the nucleus is small and flat and does not produce a bulge in the cell.

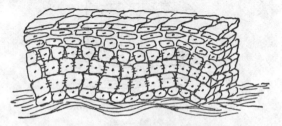

Fig. 26. Stratified Squamous Epithelium

The outer layers of cells on the body become hardened or cornified; appropriately, they are called the **stratum corneum.** This layer bears hair and nails in man, and helps form scales and hoofs in other animals by a process called **keratinization.** Hardening provides considerable protection for the animal. It may be induced by dehydration of the cells or by lack of nutrition due to the distance from the basement membrane, with the result that the outermost cells are continually sloughing off and being replaced by those underneath. The cosmetic industry capitalizes on this procedure by providing a wide variety of "moisturizing" creams designed to hydrate the dead or drying cornified cells and rejuvenate "aging" skin.

Glandular Epithelium. There are numerous categories into which the various forms of glandular epithelium may be classified. One fundamental way is simply to divide glandular epithelium into the types of glands involved. The **exocrine** type are glands with tubules that give off external secretion, and the **endocrine** type are glands of internal secretion, popularly known as the ductless glands. We then subdivide our first type, the exocrine glands, into two groups: the unicellular glands and the multicellular glands.

Unicellular exocrine glands are the single-cell glands that are scattered among the cells making up the mucosa—the tissue that lines the tubes of the trachea, intestine, and uterus. The classic example taken from man is the goblet (or mucous) cell (Fig. 27) that is prominent in the intestine; it is found also in the respiratory ducts. This cell is essentially a columnar cell that contains a large vac-

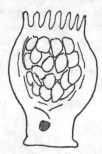

Fig. 27. Unicellular Gland (Goblet Cell)

uole of the substance called mucigen. Mucigen is the precursor of mucin, or mucus as it is sometimes called. The mucigen vacuole or mucous plug frequently takes on the shape of a long-stemmed goblet; hence the name goblet cell. The mucigen is discharged either explosively or gradually. Following a discharge, the cell reverts temporarily to a period during which it looks like an ordinary columnar cell; then it begins to accumulate mucigen within its cytoplasm and repeats the cycle.

Fig. 28. Goblet Cells

Multicellular exocrine glands are common in occurrence and diverse in type and morphology. Our examples will be drawn from those glands that are found in man. Basing our classification on the type of duct that is present, multicellular exocrine glands are regarded as either simple or compound. The simple types are those which possess an unbranched duct system; the compound are those provided with branching ducts.

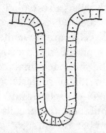

Fig. 29. Simple Tubular Gland

The **simple exocrine glands** will be briefly considered first. In man, we find **simple straight tubular** glands (Fig. 29) that are merely short straight hollow ducts lined with secretory epithelial cells. This type of gland usually opens upon a mucous membrane; for example, the intestinal glands lubricate the lining of the intestine, a mucous membrane. (The mucosa is an epithelial tissue.) We also find **simple coiled tubular** glands (Fig. 30), a duct system without branches, coiled like a short spring. This type is well exemplified by the sweat glands in man's skin.

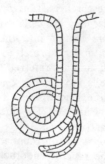

Fig. 30. Simple Coiled Tubular Gland

A third type of simple gland has a deceptive name—the **simple branched tubular** gland (Fig. 31).

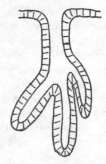

Fig. 31. Simple Branched Tubular Gland

Exemplified by the glands in the lining of the uterus, these structures have unbranched tubes, although frequently several tubes have a common opening upon the mucosa. Some simple glands are expanded and become pouchlike. These we call **saccular** (Fig. 32). The best examples are those that make up the oil glands associated with the hair follicles in the skin. Where ducts are present usually only the innermost epithelial cells are secretory; the other cells are modified to form the tubule.

Fig. 32. *Left*–Simple Saccular Gland
Right–Simple Branched Saccular Gland

The **compound exocrine glands** (Fig. 33) are characterized by a large number of ducts that branch

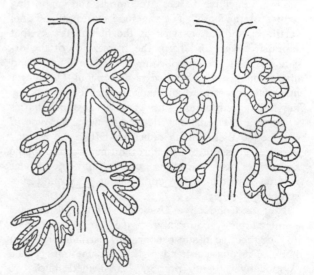

Fig. 33. *Left*–Compound Tubular Gland
Right–Compound Saccular Gland

and merge into a common main excretory duct. Most of the compound glands possess either mucous cells or serous cells which secrete a protein, albumin (like egg white); some glands have combinations of both types of cells. The major salivary glands of man demonstrate this condition, with the parenchymal cells (secretory epithelium) of some glands being entirely serous and others possessing a mixture of serous cells and mucous cells. Other compound glands are the mammary or milk glands, and the liver, the largest of all glands.

The **endocrine glands,** the so-called ductless glands, release their secretory products directly into the bloodstream. There are numerous endocrine glands in man, and the parenchymal cells of the various endocrine glands are specialized to a high degree. Recently, histochemical studies have discovered several new cell types in certain endocrine glands. These cells and their secretions, called **hormones,** are the subject of endocrinology, a large field of study in itself. The one thing that all of the endocrine glands have in common is a good blood vascular system, with their parenchymal secretory cells abutting it.

CONNECTIVE OR SUPPORTIVE TISSUES

Morphologically, connective tissues differ from the epithelial tissues in that they possess fewer cells and a relatively greater amount of intercellular material in a comparable quantity of tissue. The intercellular material varies considerably in quantity and fluidity. With rare exception, it is the intercellular material that characterizes the connective tissues.

The connective tissues are literally the supporting tissues of the body. They include the familiar bone, cartilage, and fat, as well as the fibrous tissues that support, cover, and form the framework of the organs. All of the supporting tissues have cells and fibers which are suspended in amorphous (formless) material, the ground substance, which is secreted by the cells themselves. The ground substance varies in form and consistency depending upon the type of connective tissue being considered. The consistency varies from semifluid to hard and mineralized. Some authorities use the term **matrix,** or **intercellular matrix,** to indicate the combination of ground substance and fibers.

Fibrous Connective Tissues. From the histological point of view, connective tissues are divided into: (a) connective tissue proper, (b) cartilage, and (c) bone. **Fibrous connective tissue,** sometimes called "connective tissue proper," is divided into two groups: loose connective tissue and dense connective tissue. These two groups include: the **fascia,**

the supportive layer underlying the skin; the **stroma,** or supportive fibers; **tendons; ligaments;** and **fat.** Some connective tissues possess a wide variety of cells and fibers suspended in the ground substance, while others are more limited in variety.

Basic Connective Tissue Cells. The basic cell of connective tissue proper is the **fibroblast** or **fibrocyte.** The cells so designated are large in size, varying in morphology from an elongated spindle shape to a branching stellate (star) shape. The cytoplasm is smooth and homogeneous, and it stains rather poorly with the ordinary aniline dyes. The nucleus of a fibroblast is ovoid, with fine chromatin granules that stain with only moderate intensity. Some nuclei display one or two nucleoli. The fibroblasts are active in the formation of the connective tissues and also may participate in the production of the glycoprotein of the ground substance.

Undifferentiated cells (also called mesenchymal or indifferent cells) are difficult to distinguish from fibroblasts. These cells are juvenile cells with the potential to form all of the various cells of connective tissue. The mesenchymal cells, which tend to be smaller than the fibroblasts, are stubbily stellate and have a pale cytoplasm and an oval nucleus. With proper stimulation these cells can divide and mature into a number of different cell types including the fibroblast.

Fat cells occur in varying numbers in connective tissues. The mature fat cell accumulates a large droplet of fat within its cytoplasm that displaces both the nucleus and the cytoplasm toward the edge of the cell. Most routine histological methods for preparing tissues for staining use fat solvents. Thus, the fat cell after such treatment shows a large vacuole where the fat dissolved, a thin outer rim of cytoplasm, and a peripherally placed nucleus. Such a fat cell resembles a signet ring.

Other Connective Tissue Cells. A number of less well-known connective tissue cells occur. Some of them have unusual abilities and are currently being investigated by researchers. **Macrophages** are phagocytic cells; that is, they are cells capable of ingesting minute particulate matter into their cytoplasm. In connective tissue under normal conditions they are apparently sessile, and are referred to as **fixed macrophages.** These cells are smaller than the fibroblasts, irregular in outline, and possess a vacuolated cytoplasm and an oval nucleus. The nucleus stains darker than the fibroblast nucleus.

The macrophages of the connective tissue are a component of the reticulo-endothelial system of the body, a phagocytic system that removes cellular debris, bacteria, and fine particles of matter from tissue fluid and blood. Relatively little emphasis has been placed on this system in man as compared to the commonly known systems, such as circulation and respiration. It is interesting to recall the "eating" capabilities of some free living protists for comparison.

Mast cells occur in connective tissues in variable numbers. The peculiar name, chosen by German workers who first studied the cells, means "overfed" —mast cells literally bulge with secretory granules. They are found with greatest frequency in association with small blood vessels. The cells are large but nevertheless require special staining methods to render them clearly visible. Each mast cell has a small, poor-staining nucleus, and cytoplasm that is filled with coarse granules. Recently, these cells have been researched increasingly. It has been demonstrated that they contain heparin, histamine, and serotonin—vital chemicals in the defense system of the body against wounds and other trauma.

Plasma cells are unusual cells which appear in the connective tissue. They are probably formed from the lymphocytes, which are a type of white blood cell present in normal blood. Under normal conditions plasma cells are rare, but they appear in abundance in the connective tissue in areas of chronic inflammation. The plasma cell is moderate in size, rounded or circular in sections. The cytoplasm is lightly basophilic, meaning that it stains blue when aniline dyes such as hematoxylin are applied to it. In the plasma cell the nucleus is usually eccentric in position, and its chromatin material is arranged at the periphery near the nuclear membrane like the numbers on the dial of a clock. The plasma cell is thought to be involved in the process of producing antibodies, the mechanisms of defense against foreign invaders of the body.

Other cells are found within the connective tissue although they do not originate there. Lymphocytes from circulating blood gain access to the connective tissue, as do other white blood cells (leucocytes). These cells will be described when blood is discussed as a tissue.

Fibers. The fibers of connective tissue are arranged into three types: collagenous, reticular, and elastic. The most common is the collagenous fiber, which is actually a bundle of small fibrillae cemented together into what appears to be a single unit. These collagenous fiber bundles resist stretching and impart strength to the connective tissue.

In routine histological preparations, an aniline dye, eosin, stains collagen fibers a bright pink color, while aniline blue dye renders collagen a brilliant blue. The abundance of collagen fibers and their staining properties make them easily recognized in histological sections. Electron microscopy has revealed minute cross-striations that are about 640 Angstroms apart on the collagen fibrillae.

Reticular fibers, in contrast to collagenous fibers, are exceedingly delicate, branching, threadlike strands that are found in connective tissues. There appear to be at least two types of reticular fibers: (a) precollagenous, which mature into collagenous fibers; and (b) stable reticular, which form the framework of lymphoid organs, the fine stroma of endocrine glands, support for capillaries, and part of the basement membrane supporting epithelial cells.

Reticular fibers have only a fair affinity for acid aniline dyes but a strong affinity for heavy metal staining techniques utilizing silver nitrate or gold chloride. For this reason, the term "argyrophilic" is applied to reticular fibers. This affinity is not shared by the collagenous fiber bundles. Electron microscopy has revealed the same distance of 640 Å separating cross-striae in reticular fibers as was seen in collagenous fibers.

The third type of fiber, the elastic fiber, has little tendency to form heavy bundles; it occurs instead in elongated branching fibers scattered in variable quantities among the collagenous fiber bundles. These fibers are resilient but they lack the strength of collagen. They are chemically distinct and can be stained so as to be distinguished from collagen. Of the three fiber types, less is known about elastic fibers than about the other two.

Ground Substance. The ground substance, or substrate of connective tissue proper, is a semisolid jelly-like material which appears homogeneous in the fresh state, and stains very poorly, if at all, in routine histological preparations. It is an amorphous substance that consists of a number of complex proteins and mucopolysaccharides.

As we have already seen, there are two general categories of connective tissue proper: loose connective tissue and dense connective tissue. These tissues are composed of varying proportions of the basic cell types discussed above.

Loose Connective Tissues. These are subdivided into three types of tissue. Reticular tissue (Fig. 34), our first subdivision, makes up the delicate stroma of the lymph organs, bone marrow, and the exocrine

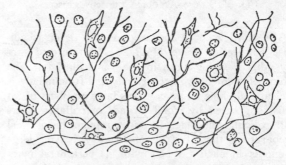

Fig. 34. Reticular Tissue

and endocrine glands. It is also associated with blood vessels and adipose tissue (fat). The cells of reticular tissue have special names: primitive reticular cells and phagocytic reticular cells.

The **primitive reticular cell** is apparently a special type of fibroblast capable of developing into a number of different types of cells. It is large and stellate, with a pale cytoplasm and a large, poorly staining nucleus. The **phagocytic reticular cells** are actually fixed macrophages that are part of the reticulo-endothelial system. The delicate reticular fibrillae form a fine network that is difficult to discern without special stains.

Loose areolar connective tissue (Fig. 35), also called simply areolar tissue, is the second type of loose connective tissue. It is the most abundant of all the connective tissue. There is hardly a microscopic preparation of animal material that one could make that would not have some of this type of tissue present. Loose areolar connective tissue contains all the cell types, the fibers, and the ground substance that have been mentioned. Its main cell type is, of course, the fibroblast. It may contain many macrophages. The other cell types, such as the mast and plasma cells, are found in fewer numbers. In the areolar tissue, there is an abundance of collagen fibers arranged in a loose meshwork containing scattered elastic fibers and some reticular fibrillae.

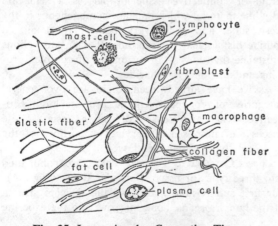

Fig. 35. Loose Areolar Connective Tissue

The third type of loose connective tissue is **adipose connective tissue** (Fig. 36), also termed simply adipose tissue, and, in the vernacular, fat. The major element here histologically and functionally is the fat cell. This tissue contains large numbers of fat cells, supported by small strands of collagen fibers and reticular fibrillae. In most routine histological preparations the fat is dissolved away, leaving large vacuoles in each fat cell with its peripheral cytoplasm and nucleus. If you visualize chicken wire, you can get an approximation of the appearance of this tissue on slides.

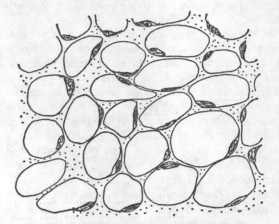

Fig. 36. Adipose Connective Tissue

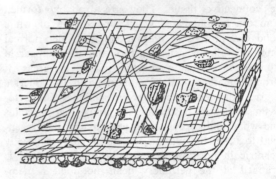

Fig. 37. Dense Interlacing Connective Tissue

Dense Connective Tissues. These are characterized by an abundance of collagenous fiber bundles

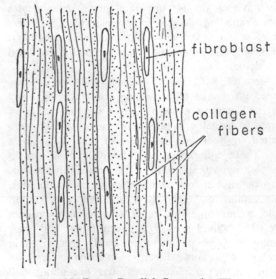

Fig. 38. Dense Parallel Connective Tissue

and a proportional reduction in cells. This category of connective tissue is subdivided into dense interlacing connective tissue and dense parallel connective tissue. The **dense interlacing connective tissue** (Fig. 37) has irregular interwoven collagen fiber bundles and scattered fibroblasts. This form is seen, for example, in capsules that form around some of the organs and in the sheath covering the bone. **Dense parallel connective tissue** (Fig. 38) has regularly arranged, compacted collagen fiber bundles, among which are interspersed scattered fibroblasts. This is the principal type of tissue which forms tendons.

Cartilage. Cartilage is highly specialized connective tissue. Like the connective tissues proper, discussed earlier, it also is composed of cells, fibers, and a ground substance. Cartilage, the familiar gristle of the body, is firm, flexible, resilient, and will bear moderate amounts of weight. In cartilage, there is only one specialized type of cell, the **chondrocyte.** The chondrocyte is large and spherical, with basophilic cytoplasm and a prominent, moderate-staining nucleus. Each chondrocyte lies within a space in the ground substance called a **lacuna** (*pl.:* lacunae). Collectively, the ground substance and the collagen fibers constitute the **cartilage matrix.** In routine histological preparations the cartilage matrix is basophilic; the collagen fibers blend into the ground substance and are not visible unless special techniques are used to reveal them.

There are three types of cartilage from the histological point of view: hyaline, elastic, and fibrous. **Hyaline cartilage** (Fig. 39) is the most abundant of the cartilages; it occurs in sheets, plates, or irregular masses. In man, it is found in the nose, larynx ("voice box"), upper respiratory ducts (trachea and large bronchi), at the ventral tips of the ribs, and in the articulating ends of the long bones (at the joints of arms and legs). With the exception of the cartilage at the articular surfaces, the cartilage areas are enveloped in a layer of dense interlacing connective tissue that forms a protective covering called the **perichondrium.**

Elastic cartilage (Fig. 40) is found in the external shell (auricle) of the ear, and in the throat as the epiglottis and part of the larynx; in these areas flexibility is a prime necessity. Histologically, elastic cartilage is like hyaline cartilage, with the addition of elastic fibers in the cartilage matrix. It has chondrocytes in lacunae and is covered with perichondrium. With special stains, the elastic fibers can be made visible.

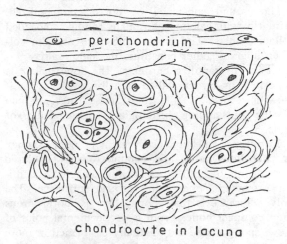

Fig. 40. Elastic Cartilage

Fibrous cartilage (or fibrocartilage) (Fig. 41) is found in areas where strength is required, such as in the discs that lie between the vertebrae in the spine. Fibrocartilage is composed of chondrocytes in lacunae and the matrix contains a greatly increased

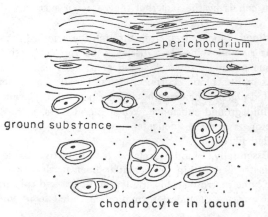

Fig. 39. Hyaline Cartilage

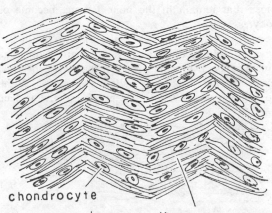

Fig. 41. Fibrous Cartilage

number of collagen fiber bundles. These bundles give the tissue a fibrous appearance. It has no perichondrium.

Osseous Tissue. Bone, sometimes called osseous tissue, shows the greatest degree of specialization of all the supporting tissues. It is, however, like all the connective tissues, endowed with cells, fibers, and a ground substance secreted by the cells. Mature bone cells are called **osteocytes** and are located within lacunae in the bone matrix. Osteocytes are large and irregularly oval, with the cytoplasm radiating out into minute channels in the matrix called **canaliculi** (little canals). Each osteocyte is provided with a large, moderately basophylic nucleus.

The ground substance of bone is mineralized, primarily with calcium phosphate, and has embedded collagen fibers. The ground substance and the collagen fibers together constitute the **bone matrix,** giving bone its familiar hardness. The bone matrix is deposited in layers, called **lamellae,** which give it the appearance of tree rings; these rings are one of the outstanding histological characteristics of bone.

Two types of bone are present in man: **spongy** or cancellous bone, and **compact** bone. The basic components of both types are the same, with identical osteocytes, fibers, and ground substance. The only difference is that spongy bone is more porous than compact bone. From a developmental point of view, spongy bone may actually be formed in more than one way. It may be membranous, formed by deposition of mineral over fibers, or it may be cartilage bone. In the latter case, cartilage forms first, and then is eroded by osteoclasts until only a framework remains. Spongy bone then forms by calcium deposition. Compact bone is formed by filling in the framework with more mineral deposits, and by crowding. Osteoblast cells form the developing bone and most of these move out as the lamellae are laid down. Those that remain in the bone lacunae become osteocytes.

Spongy bone (Fig. 42) is the simpler of the two types, therefore, and has a very irregular organization. The bone lamellae form a meshwork of small beams or bars of bone called **trabeculae.** Within the trabeculae the osteocytes are found in the lacu-

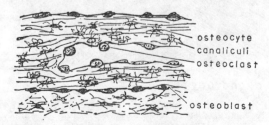

Fig. 42. Spongy Bone

nae. The relatively large spaces between trabeculae are occupied by connective tissues—bone marrow or fat. Obviously, this is an extremely porous form of bone.

Compact bone (Fig. 43) forms the shafts of long bones which typically have large marrow-filled central cavities. Around the outer surface of bone is a dense interlacing connective tissue sheath called the **periosteum.** A similar sheath, the **endosteum,** lines the marrow cavity.

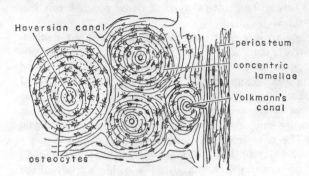

Fig. 43. Compact Bone

Compact bone is characterized by the orderly arrangement of the lamellae. Inside the periosteum and endosteum are a series of parallel lamellae. The bulk of the compact bone has lamellae arranged in units termed **Haversian systems.** The central portion of each Haversian system is a canal containing small blood vessels; this is called the **Haversian canal.** Around each Haversian canal there are from five to forty bone lamellae arranged in concentric circles; the bone between Haversian systems is composed of irregular lamellae. Smaller channels containing blood vessels penetrate compact bone and enter the Haversian canals almost at right angles. These are termed **Volkmann's canals,** or communicating canals, and are readily distinguished from the Haversian canals by the fact that they are not surrounded by concentric lamellae.

Blood. Blood, circulating through the arteries, capillaries, and veins, is a tissue. Like the connective tissue already discussed, blood consists of cells (red and white blood corpuscles) and cell fragments called **platelets** or **thrombocytes** suspended in **plasma,** a fluid ground substance. When blood clots, a third factor is present, delicate threads of a substance called **fibrin.**

Essentially, blood is related to connective tissue. It is formed within the reticular fiber meshwork of the bone marrow and in the lymph nodes and spleen. The process of blood formation is called **hemopoiesis.**

Blood plasma, the intercellular matrix of blood,

is histologically an amorphous, homogeneous substance. Chemically, it contains protein, carbohydrates, lipids, and inorganic salts. It also contains dissolved hormones and enzymes. Physiologically, plasma suspends the blood corpuscles, serves as a vehicle for enzymes, hormones, and digested nutrients, and contains the substances which enter into the formation of the fibrin strands when blood clots. On slides plasma appears histologically as a fine, granular, pink-staining, structureless precipitate.

Suspended in the plasma are the formed elements of the blood, the red blood corpuscles (erythrocytes), white blood corpuscles (leucocytes), and the blood platelets (thrombocytes).

The normal erythrocytes (Fig. 44) in man and other mammals are nonmotile cells lacking nuclei. In the lower vertebrates (reptiles, birds, amphibians, fishes) the erythrocytes possess nuclei. In man, the erythrocytes contain hemoglobin which imparts the red color to circulating blood. The normal adult woman has about 5 million erythrocytes per cubic millimeter of blood; the normal adult man has about 5.5 million erythrocytes per cubic millimeter of blood. Morphologically, in humans, the erythrocytes are small (average diameter in fixed tissue is 8 microns) biconcave discs of uniform size. The erythrocytes stain a pale pink with eosin dye.

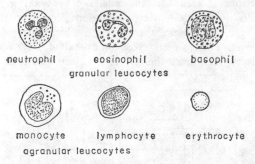

neutrophil   eosinophil   basophil
granular leucocytes

monocyte   lymphocyte   erythrocyte
agranular leucocytes

Fig. 44. Blood Cells

The white blood corpuscles are collectively termed leucocytes. All of the leucocytes are true cells, containing nuclei, and are capable of some independent movement. In man, there are five types of leucocytes divided into two major subgroups: the granulocytes (granular leucocytes) and the agranulocytes (agranular leucocytes). The granulocytes include the neutrophil, the eosinphil, and the basophil, all having characteristic granules in their cytoplasm which stain lavender, red, and dark blue respectively. The agranulocytes are the lymphocytes and the monocytes which have poorly defined or no cytoplasmic granulation. See Fig. 44.

The neutrophil is the most abundant granular leucocyte, furnishing 60 to 70 per cent of the total leucocytes. It ranges from 12 to 15 microns in diameter. Its dark-staining nucleus is lobed, sometimes elongated, and usually consists of two or three irregular oval lobes. The cytoplasm contains numerous well-defined, lightly acidophilic granules.

The eosinophil occurs infrequently under normal conditions, constituting about 3 per cent of the leucocytes. It is about the same size as the neutrophil and has an elongated or lobed nucleus. The cytoplasm is filled with coarse granules that are strongly acidophilic in their staining reaction, producing a bright red color with eosin dye.

The basophil is the rarest of all the leucocytes, amounting to only 0.5 per cent. The basophil is the same size as the neutrophil and the eosinophil, and also contains a lobulate nucleus. The numerous coarse cytoplasmic granules stain a dark blue.

The lymphocytes are fairly abundant agranular leucocytes, making up from 25 per cent to 35 per cent of the total number of leucocytes. Their size is from 10 to 12 microns in diameter. The lymphocytes have a large nucleus, oval to spherical and dark-staining, which occupies much of the cell. The cytoplasm is homogeneous, lightly basophilic, and forms a relatively narrow rim around the nucleus in the stained and fixed material.

The monocyte, also an agranular leucocyte, makes up about 5 per cent of the total white blood cell population. It is about 15 microns in diameter and has an eccentrically placed nucleus that may be kidney-shaped. The cytoplasm is relatively abundant, lightly basophilic, and contains variable numbers of scattered fine granules.

The blood platelets (thrombocytes) are cell fragments, not true cells. They are very small, ranging from 2 to 4 microns in diameter. There are 300,000 thrombocytes per cubic millimeter in the circulating blood. These minute structures are basophilic in staining properties and have no uniform shape. They are important in blood coagulation.

Muscle. The various cells of muscle have two things in common: they undergo contraction and they are elongate in form. The physiologist classifies muscle into two categories: **voluntary** muscle (under control of the will) and **involuntary** muscle (not under control of the will). The morphological classification of muscle recognizes two categories—**smooth** muscle and **striated** muscle. It so happens that most of the striated muscle is voluntary with one exception—the cardiac muscle of the heart. Almost all of the smooth muscle is involuntary. A useful classification is:

Smooth: Involuntary
Striated: Voluntary (skeletal)
Striated: Involuntary (cardiac)

Smooth muscle, histologically, is composed of cells that are elongated, tapering (spindle-shaped), and cylindrical in form. Their size varies with location: in the wall of a small blood vessel smooth muscle cells may be 12 to 15 microns long; in the muscular layer of the intestine they may be 100 microns long; in the uterus of the pregnant woman they may be up to 500 microns long.

The cytoplasm of the smooth muscle cell (Fig. 45) consists of two structural elements. The **myofibrillae** are threadlike and are longitudinally arranged parallel to the long axis of the muscle cell. They are, in turn, suspended in the other element, **sarcoplasm,** which is the fluid constituent of the muscle cell. The nucleus of the smooth muscle cell is centrally located in the widest part of the tapering cell. It is elliptical, ovoid, and stains with moderate intensity. Most smooth muscle is organized so that the cells are closely arranged in sheets or bundles which makes the determination of individual cell outlines somewhat difficult.

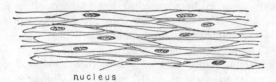

Fig. 45. Smooth Muscle

It was noted earlier that there are two types of striated muscle: striated skeletal muscle (Fig. 46) and striated cardiac muscle (Fig. 47). Individual skeletal muscle cells are elongated, ribbon-like. These cells may be 1000 microns or more in length and up to 100 microns wide. They are normally found gathered together in bundles of varying size, supported by fibrous connective tissue. These cells contain a comparatively large amount of cytoplasm (sarcoplasm) and as a result we find that the skeletal muscle cells are multinucleate. The nuclei are numerous and are located in the peripheral cytoplasm immediately inside the cell membrane. In man, this peripheral location is an important way by which skeletal muscle can be distinguished from cardiac muscle. Each skeletal muscle cell is enclosed in a thin, almost transparent, membrane called the **sarcolemma.** Within the cell there are longitudinally dispersed myofibrillae suspended in the sarcoplasm. Striated muscle cells (both skeletal and cardiac) exhibit regular cross-banding or striations. This is because the myofibrillae are composed of regularly alternating light and dark material. In skeletal muscle the cross-striations are very well marked. The dark

bands are termed the "A" discs and the light striations are called the "I" discs. In some preparations, the light band is bisected by a thin dark line that is known as the "Z" line. ("A"=anisotropic; "I"=isotropic; "Z"=zwischenscheibe or "disc between.")

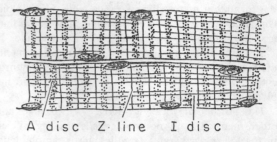

Fig. 46. Striated Skeletal Muscle

Cardiac muscle, the muscle of the heart, bears some resemblance to skeletal muscle: it is elongated, ribbon-like, and has cross-striations. In man, however, it differs in a number of important respects. In cardiac muscle, the muscle cells tend to branch or divide, a tendency that does not exist in skeletal muscle. Also, in cardiac muscle the nuclei are centrally placed in the sarcoplasm, rather than in the peripheral area like the nuclei of skeletal muscle cells. In addition to cross-striations similar to those in skeletal muscle, there are prominent, irregularly placed and formed cross-bands called **intercalated discs** which mark the terminus of adjacent cells.

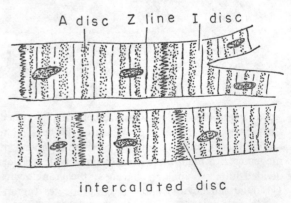

Fig. 47. Striated Cardiac Muscle

CELLS OF THE NERVOUS SYSTEM

In treating the histology of the nervous system, the discussion will be limited to the structural ele-

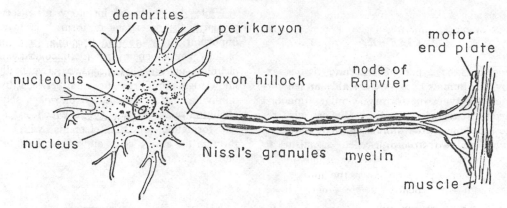

Fig. 48. Motor Neuron

ments of the central nervous system, which consists of the brain and spinal cord, and of the peripheral nervous system, which is comprised of the spinal and cranial (brain) nerves.

The **central nervous system** structures have two areas, termed the **gray matter** and the **white matter** because of their respective appearances. The gray matter contains the nerve cells and their extensions, called **dendrites** and **axons**, plus the supportive cells called **glia cells** or **neuroglia.** The white matter is composed of bundles of nerve fibers enveloped in sheaths of a lipid-protein mixture called **myelin.**

Each nerve cell or neuron, histologically, is divided into two regions (Fig. 48). The cell body, or **perikaryon,** contains a distinct spherical nucleus with a prominent nucleolus, and the cytoplasm (neuroplasm) contains minute filaments called neurofibrillae and a granular material called the chromophil substance (or Nissl's granules). These granules are believed to be ribosomes lying along the endoplasmic reticulum. The nerve cell extensions consist of one, or, usually, several dendrites (dendrons) and a single axon. The dendrites are comparatively short, branching, direct extensions of the nerve cell body. The axon frequently is extremely long and slender, and arises from an area of the nerve cell body called the axon hillock. The axon is commonly called a nerve fiber.

The second type of cell in the central nervous system, the glia or neuroglia, is actually a collection of cells that fall into at least three morphological categories: astrocytes, oligodendroglia, and microglia. The astrocytes are stellate with many thin radiating cytoplasmic extensions. Some authors credit these cells with a supportive function, others with an insulating function. The oligodendroglia are smaller, with few short protoplasmic extensions. Their function has not been definitely established. The mi-

croglia, the smallest of the three types, are ovoid and are believed to be phagocytic in function. See Fig. 111.

The peripheral nervous system contains the cranial and the spinal nerves. Each "nerve" is composed of a large number of nerve fibers, the term "fiber" referring to axons and their covering sheaths. Histologically, the axon appears as a thin fibrillar threadlike structure of uniform thickness as seen in longitudinal sections.

Certain dyes, such as methylene blue, reveal delicate neurofibrillae in the axon. Enveloping the axon is a delicate myelin sheath. During the preparation of slides, the lipid material is dissolved by the solutions routinely used. As a result, the myelin sheath is pale in staining and shows numerous irregular vacuolated areas.

Directly outside the myelin sheath in the peripheral nerves is found the Schwann sheath, also known as the neurolemma, a thin cellular sheath showing, in longitudinal sections, a thin peripheral cytoplasm and scattered elongated nuclei (Fig. 49). Each neurolemma cell surrounds a short segment of the myelinated axon and between adjacent neurolemma cells, where two cells meet, there is a thin but definite constriction that is called the **node of Ranvier.** It is believed that the neurolemma cells secrete the noncellular myelin coating.

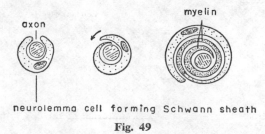

neurolemma cell forming Schwann sheath

Fig. 49

CONCLUSION

In reviewing the types of cells and tissues exhibited by the human body, we are able to see certain basic cell types performing similar functions in the several categories of tissues. We can also relate these functions to the basic equipment contained in all living cells—from single-celled organisms to man.

While the study of histology may seem to be primarily the describing of form, it is an extremely important study now that cellular and molecular biology and biochemistry are uncovering the basic chemical units of living systems. Eventually, man's knowledge may draw no line between living and nonliving matter. Instead, an unbroken sequence of evolution may be uncovered which extends from the formation of the elements with hydrogen, through the amino acids and nucleotides, to protists, and to man.

CHAPTER 5

REPRODUCTION AT THE CELLULAR LEVEL

Who was the first person to observe cell division and to comprehend its significance? This question cannot be answered by naming one man. Several early scientists apparently observed the presence of cells in plants and in animals; the question is whether they considered cells to be the essential structural unit, or mere curiosities. Many of the scientists gradually came to a realization of the importance of the cell. Sometimes, in considering the writings of early scientists, it is easy to presume that present-day knowledge beyond their grasp exists in their works. This is like interpreting the fortune-teller's predictions—the weight of matching a recent event to an earlier cryptic prediction falls on the person who makes the association.

The German botanist Matthew Schleiden took note of earlier descriptions of the nucleus in plant cells and also remarked on the presence of the nucleus in even the youngest embryo cells. Schleiden concluded that the nucleus must hold some close relationship to the development of the cell. He mistakenly concluded that a new cell arose by budding from the nucleus itself.

Theodor Schwann, the German naturalist, in 1839 suggested that the eggs of animals were also cells because they contained all the required parts, such as nucleus and cell membrane, no matter how distended the egg might be with food substances. The discovery of the tiny mammalian egg in 1828 by the embryologist Karl Ernst von Baer probably influenced Schwann because the mammalian egg is microscopic and has almost no food, while the hen's egg is loaded with food material.

The zoologists Karl Theodor von Siebold and Michael Sars in 1837 had observed cell division in invertebrates and, in 1838, "segmentation" or cleavage of the mammalian egg was described. Schwann saw the cleavage of a hen's egg in 1839 and, in 1841, the division of frog's eggs was first described, although many earlier workers had studied this phenomenon.

Around 1842–46, the Swiss botanist Karl von Nägeli, by studying cell division in lower plants, soon found Schwann's idea of cells budding from the nucleus to be erroneous. Franz Leydig, the German zoologist, and Robert Remak, the German physiolo-gist, were the first to perceive adequately, in 1848 and 1852 respectively, the division of nuclei in the cleavage of ova. Leydig and Remak were colleagues of the Swiss anatomist Rudolf Albert von Kölliker. These scientists were impressed with the importance of nuclear division; von Kölliker must have been one of the first to regard the nucleus as the transmitter of hereditary characteristics.

With the discarding of the notion that the new cell budded from the old nucleus, von Kölliker and the others first believed that the nucleus simply disappeared during cell division and reappeared in the offspring daughter cells. Improved techniques in staining cells in the 1870s made it apparent that the nucleus did not disappear, but underwent a precise series of changes.

Again it can be seen that a certain level of technique must be reached and a certain climate of opinion must be present for a new development in knowledge to occur. At this point, the theory that the cell was the unit of life was fairly well established.

The famous words of William Harvey in 1651, "ex ovo omnia"—all [creatures] from the egg—had been joined by the now equally famous words of Rudolf Virchow, "omnis cellula e cellula"—all cells from cells—stated in 1858. Virchow was a brilliant editor and professor at Berlin who studied disease and its relation to tissues. He began the study of pathology and much of his work is still valid. Virchow referred to the body as a state in which every cell is a citizen. Louis Pasteur, in the process of demolishing the theory of spontaneous generation in 1859 and the succeeding few years, added the famous words "omne vivum e vivo"—every living thing from a living thing.

So a number of eager investigators in a sympathetic atmosphere began to observe the complex process of cell division. Eduard Strasburger, a botany professor at Bonn, wrote a volume on cell formation and cell division in plants in 1875 that was published in several editions. In 1882, Walther Flemming wrote the pioneer study of cell division in animals. These early descriptions are for the most part still in use, and the vocabulary the two men used has become the accepted terminology in the field. Flemming is considered the father of modern cytology—the study of cells.

MITOSIS—LIFE FROM LIFE

Flemming introduced the term **mitosis** (from the Greek word for "thread") as the name for the process of division in the cell. The word "protoplasm" for the watery cell content material had been coined by the botanist Hugo von Mohl in 1846. Now Strasburger divided the protoplasm into two parts: nucleoplasm, the material inside the nucleus; and cytoplasm, the material in the cell outside the nucleus. Since plant cells have nonliving cell walls, the "plasms" were considered to be the living portions of the cells.

Strasburger introduced the names for three of the successive stages of cell division: **prophase,** meaning the appearance before; **metaphase,** the appearance between or after; and **anaphase,** the upward phase

or appearance. The final stage is known as **telophase,** the end phase.

Mitosis is the orderly process for the separation of duplicates, and the transmission of these to separate cells. The material that is duplicated is the **chromatin**—dark-staining threadlike strands within the nucleus which contain the genetic code.

PHASES OF MITOSIS

Prophase. The first indication that a cell is about to divide is seen when the nuclear reticulum of chromatin begins to condense into long, thin, stainable strands called **chromonema** or **chromonemata** (*nema*=thread). This is the beginning of the prophase stage. See Fig. 50.

Duplication of the strands may occur during early prophase, but it is more commonly accomplished prior to the beginning of mitosis. The threadlike double strands become coiled and twisted upon each other, forming a double spiral or helix. Each

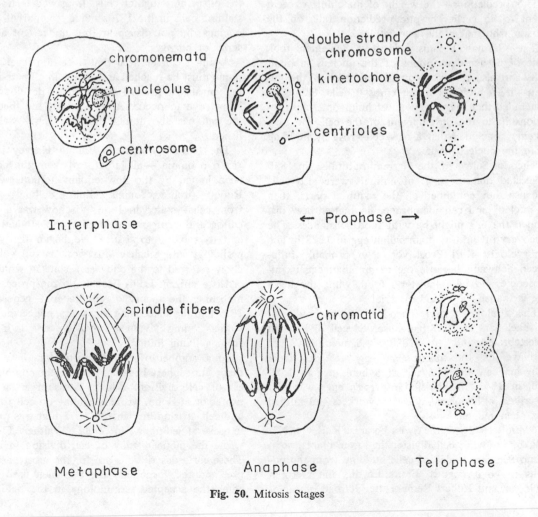

Fig. 50. Mitosis Stages

double strand is held together by its activity center (**centromere** or **kinetochore**). The chromonemata gradually condense until they are short and thick, assuming the rodlike shape characteristic of chromosomes.

The nucleolus gradually disappears during this time. Some authorities believe that the RNA contained in the nucleolus is added to the chromosomes, while others believe it is released into the cytoplasm when the nuclear membrane disappears. This takes place gradually during prophase.

At the same time that the nuclear reticulum begins to condense, another activity is taking place outside the nucleus. The centrosome body divides, releasing the two centrioles. The centrioles, you remember, have been related by some authorities to the activity centers seen at the base of the flagella in some of the protists and may be one indication of the carry-over from unicellular structure into the multicellular animals. The two centrioles now separate and migrate toward opposite ends of the cell.

Next, fibers begin to form around each centriole in a star shape, and long fibers connect the two centrioles to form a spindle shape, while shorter fibers extend only to the middle of the spindle. If the centrioles are thought of as being the north and south poles, the spindle fibers look like the lines of longitude on a world globe. If you have sprinkled iron filings on a paper held over a bar magnet, the pattern formed is similar to that of the spindle.

The origin of the spindle fibers is uncertain. Some authorities think that the nuclear membrane is an extension of the endoplasmic reticulum, formed into a double fold around the nucleus; when the chromosomes are ready to divide, the nuclear membrane disappears simply by unwrapping. The same reticulum may then form the spindle fibers. A somewhat different theory holds that the "nuclear sap" or **karyolymph** (or nucleoplasm), released when the nuclear membrane disappears, combines with cytoplasm to form the spindle fibers. Since we know that protoplasm may exist in both sol and gel states, the fibers would represent gel strands lying in the sol cytoplasm.

Whatever the origin of the spindle fibers may be, their formation concludes the prophase stage, the longest of the stages of cell division. Mitosis now enters the next stage.

Metaphase. Metaphase occurs when the double-strand chromosomes that have been released from their nuclear envelope line up in an orderly pattern at the equator of the spindle. The activity center of each chromosome comes to lie exactly on the midplane, and the shorter spindle fibers attach to each kinetochore. This is a short phase, and the cell soon proceeds into another short phase, anaphase.

Anaphase. The Greek root *ana-* means "upward," so it is easy to remember what happens in anaphase. The chromosomes, attached to the short spindle fibers from centriole to kinetochore, begin to divide. Each double-strand chromosome separates into two single strands; the kinetochore also divides. The kinetochore is known to have some potential for motion; in one group of snails, the kinetochores sometimes break loose from the chromosomes, grow flagella, and wander off independently.

In anaphase, the single-strand chromosomes, or **chromatids,** are propelled "up" the spindle fibers toward each centriole, separating each chromatid from its former partner. The force that propels them has not been defined. Perhaps the chromosome activity center is able to move independently along the fiber, like a car on a track, or the spindle fiber may actually pull the chromatid along, or the centriole may exert some kind of attraction. The kinetochore is attached to a short fibril and appears to drag the rest of the chromatid along forcibly. In any case, the single-strand chromatids arrive at the centrioles at both poles and the nuclear material has been divided into two equal parts—a change known as **karyokinesis.** Interzonal lines of unknown function extend between the chromatids at opposite poles until well into the final stage.

Telophase. In the final phase of mitosis, telophase, a new nuclear membrane forms around each group of chromatids. One or more nucleoli appear inside each nucleus; it is debatable whether these nucleoli represent the old nucleolus divided between the daughter cells or whether they are built by the new nucleus. The single-strand chromatids elongate and uncoil.

As the nuclei are enclosed in new membranes, the cell membrane begins to constrict midway between, until only a narrow bridge of cytoplasm links the two halves. Finally, this breaks, leaving two daughter cells where a single cell existed originally. It appears that when a cell has finally divided, organelles such as the mitochondria are also divided so that each new daughter cell has the necessary parts to begin functioning as a unit.

Interphase. Before the new cells are themselves able to divide, there is a period of rest and of active rebuilding called interphase. The period may be very short in actively growing areas, or long in relatively inactive areas. Interphase was, for years, thought to be simply a waiting period between divisions. More recent research reveals that some important activities are carried on during this time. The chromosomes are nonstaining during interphase, the coils unwind, and material is gathered to duplicate or double the strands. This phenomenon will be discussed later.

Mitosis in Plants. Mitosis in animal cells differs in only a few ways from that in plant cells. Centrioles and the aster system of rays are not present in plants. Apparently this is a development that occurred in the animal line of evolution after the split at the protistan level of living systems.

Another difference concerns the division of the cytoplasm. Since plants have nonliving cellulose cell walls, it is impossible for them to constrict in the same manner as animal cells. Instead, a cell plate forms in the cytoplasm of the plant cell at the area of the future division in the cytoplasm; this grows toward the existing side walls until the partition is complete.

Animal cell division of the cytoplasm resembles the manner in which a large droplet of water separates into two smaller ones. It stretches out and narrows the connecting bridge until the bridge suddenly parts, and the droplets round up due to surface tension. Since protoplasm is largely water, it is natural that this resemblance should be seen.

Typical mitosis in plants can be observed in onion root-tip tissue when it is stained and examined under the light microscope. Mitosis in animal cells is often observed in whitefish eggs that have undergone cell division, or cleavage. Stained slides must be made of eggs which do not contain a great deal of yolk as this would obscure the cell activity. Mitosis can sometimes be observed in living cells under the phase-contrast microscope.

The precise patterns of mitosis have been actively studied for over a century, but it was not until the advent of the electron microscope and the development of advanced techniques in histochemistry that an understanding of the changes made during mitosis was possible. It was concluded quite early, however, that the chromosomes must carry hereditary influences, since the offspring cells were duplicates of the adult that divided.

The main change in current interpretation concerns the actual division of the individual chromosomes. It appeared that the chromosome split at metaphase, being physically pulled into two separate halves, or threads. Yet, if splitting continued to occur in division after division, the chromosome would soon be fractioned into impossibly small parts.

When the chromosomes appear at prophase, they condense into normal-sized chromosomes each time. Therefore, some means of "growing" must occur in the chromosome during interphase, the stage that was at first thought to be a resting period. No increase in size can be observed before or during metaphase to support the previous idea that chromosomes double at metaphase and then split. The metabolic machinery of the cell is practically shut down during this period.

THE NATURE OF GENES AND CHROMOSOMES

The discovery of how the chromosome is able to double itself is one of the most fascinating detective stories in zoology. Before even a rudimentary answer could be given, however, it was necessary to build, slowly and methodically, a body of knowledge about chromosomes.

Even though von Kölliker, Leydig, and Remak believed that chromosomes contained the mechanism for heredity, proving this theory took time and effort. Friedrich Meischer, as early as 1869, identified the chemical composing chromosomes as nucleoprotein which he called "nuclein," but, at that time, this knowledge contributed little to the understanding of the hereditary mechanism.

Flemming and Strasburger had assumed that all chromosomes were alike. Experiments on fertilized ova by the German zoologists Theodor Boveri and Oskar Hertwig, and others indicated that the actual number of chromosomes was not the factor giving variation; consequently, it was concluded that there must be factors within the chromosomes which control hereditary characteristics.

August Weismann, the German biologist, favored a theory of "germ plasm," in which the germ cells were carried unchanged from generation to generation. Variations, he concluded, were caused by the mixing of different amounts of "ids"—little pieces of chromosomes. In 1887 Weismann also advanced the theory that, when egg and sperm meet, these "ids" doubled in number, receiving characteristics from each parent, and that some kind of splitting occurred that differed from mitosis. That this judgment was correct is shown later when the specialized cell division that occurs in the sex cells is discussed.

There is no relationship between the complexity of an animal and the number of chromosomes in each of its cells—the number ranges from two to hundreds, but the number is consistent for a given species.

A few years after Weismann's important contributions, a means of experimentation developed which has led the way in studies of chromosomes. In 1901, William W. E. Castle at Harvard had begun using a fruit fly, *Drosophila melanogaster*, in some experiments. This was a happy choice, for the little flies that dart around fermenting fruit are readily available, can be raised in milk bottles on various jelly media, reproduce rapidly, and, more luck, have only four chromosome pairs that are easy to see and distinguish from each other.

Soon, in 1906, Thomas Hunt Morgan and his colleagues at Columbia began to use *Drosophila*, as did the geneticists C. B. Bridges, H. J. Muller, and A. H. Sturtevant later. These men gradually developed a complex theory of the gene, based largely on research done with *Drosophila*.

It must be remembered that none of these men had ever actually seen the structure they called a "gene." By experimentation, scientists were able to map locations on the chromosomes that controlled certain inherited traits, such as eye color and wing length. They found, too, that when several traits were contained in the same chromosome, all of these traits were transferred at the same time. They called each single factor that controlled a trait a **gene.**

It is not surprising, then, that biologists began to think of genes as groups of beads strung on chromosomes. Yet the biochemists showed the chromosomes to be composed of nucleoprotein, and to be of uniform composition, rather than a string of differing "beads." How, then, do the genes control heredity?

As early as 1894, Hans Driesch, a German experimental biologist, had made the suggestion that heredity factors were carried out by functioning as enzymes. In order for the hereditary factors (the genes) to do their jobs, they must be able to reproduce themselves at the time of cell division, and they must also send out some sort of control substances to the cytoplasm, such as enzymes, that would direct the cells to carry out the inherited instructions.

The English physician Sir Archibald Garrod, called the "father of biochemical genetics," also had proposed that the genes produced enzymes. He had become interested in some congenital diseases which he considered to be "inborn errors in metabolism." He postulated that an error or deficiency in a given gene would mean the lack of a certain enzyme. This lack, in turn, would prevent the occurrence of the chemical step that the enzyme was supposed to catalyze. Although he published his findings in 1909 and 1923, the geneticists took no note of them since they were occupied with *Drosophila* studies. Garrod's work waited for recognition until the 1940s when biochemical techniques began to be applied to genetic studies.

George W. Beadle and Edward L. Tatum performed a long series of experiments at Stanford University and California Institute of Technology in the 1940s. Using the bread mold *Neurospora*, they showed how genes functioned as enzymes; they received a Nobel Prize in 1958 for their work. *Neurospora* is able to synthesize all the vitamins needed, except biotin, when raised on a sugar medium. By treating the spores with X rays, the gene that controlled production of another vitamin, thiamine, was destroyed.

Suppose that, in growth, molecule *A* becomes molecule *B* only when catalyzed by enzyme *b*. Enzyme *b*, however, is produced by Gene *B*. If we destroy Gene *B*, enzyme *b* is no longer produced, and molecule *A* can't become molecule *B*. See Fig. 51.

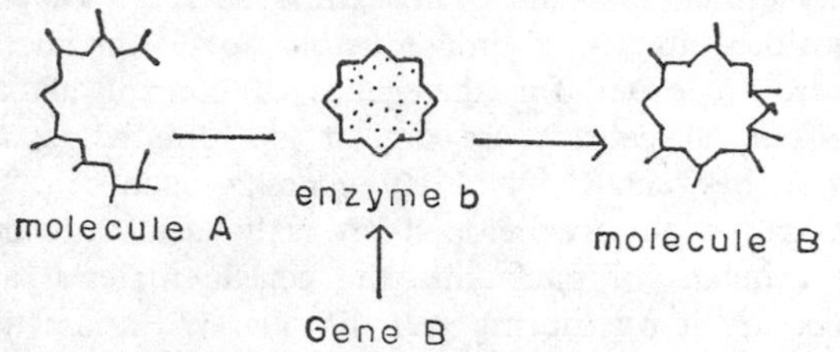

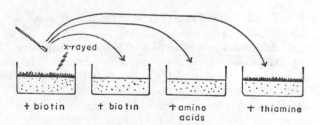

Fig. 51

This is the sort of experiment that was made with *Neurospora*. The irradiated spores were allowed to develop on a medium supplied with all necessities. Then they were transferred to various test media to discover what was lacking in the molds. The spores on a medium lacking vitamins showed no growth; those supplied with all essential amino acids showed no growth. But those grown on media to which thiamine was added showed good growth. See Fig. 52.

Fig. 52. *Neurospora* Mold Growth

Another researcher who helped demonstrate the enzyme nature of genes was P. B. Medawar, a British zoologist, who transplanted tissues and was able to show the development of antibody substances in the new site similar to those the body forms to destroy invading micro-organisms. He then applied the antibody substance to the original tissue and showed that the single gene which gave off the specific single enzyme to make thiamine could be destroyed by applying the antibody substance. Medawar received a Nobel Prize for other antigen studies in 1960.

The field of molecular biology has advanced rapidly and received increased attention during the last twenty years. A number of Nobel Prizes have been awarded to scientists making discoveries that helped unravel the secrets of genes and chromosomes.

A gene is extremely small—between 10 and 100 millimicrons in diameter—something like the smallest unit on your ruler, a sixteenth of an inch, divided by one hundred thousand. These tiny bits are composed of some 100 to 100,000 nucleotides with molecular weight of about 750. You recall that nucleotides are formed when a phosphate sugar joins either a purine or a pyrimidine ring. The nucleotides, in turn, link together to form nucleic acids. In order for the genes to control all the millions of cells, there are, it is estimated, somewhere between 10^4 and 10^7 genes in each cell.

Chromosomes developed probably as a more efficient means of packaging the genetic material and parceling it out during cell division. We know that among some of the lowest living things, the blue-green algae and some bacteria, chromatin is scattered through the cell without a nucleus or chromosomes. A pattern of evolutionary development is shown by examining the condition in which chromatin is found at various levels of organization.

Chromatin is gathered into a structure called the **endosome,** which lacks a nuclear membrane, in some higher bacteria. Next, in some of the protists, the endosome is enclosed in nucleoplasm and a membrane to form a nucleus; and in some euglenoids chromosome strands are seen. According to some authorities, the endosome eventually becomes the nucleolus in multicellular animals and higher plants. The spindle fibers are first seen in the amoeboids and ciliates.

The knowledge of the role of genes was enhanced by the period of classic *Drosophila* genetics in the first forty years of this century. However, the composition of a gene and the process by which it duplicates itself in sending out its enzyme message were still unknown.

It had been assumed that genes were proteins, since protein is found in chromosomes; also found are DNA and RNA, the "nucleins" Meischer identified in 1894. The protein theory began to be doubted, especially when O. T. Avery, C. M. MacLeod, and Maclyn McCarty at the Rockefeller Institute succeeded in extracting pure DNA from one genetic strain of pneumonia bacteria (*Pneumococcus*) and injecting it into another genetic strain of pneumonia bacteria. The pure DNA transferred genetic information from the first strain, and it replaced the genetic patterns in the second. No protein was involved in the transfer.

The work of Avery and his colleagues was based on changes in pneumococci that had been noted in 1928 by F. J. Griffith while he was studying the effects of various *Pneumococcus* strains on mice. He concluded that the bacteria changed from one kind to another, but he did not know the cause or nature of the change.

In another series of experiments, bacterial viruses (bacteriophages) became the experimental organisms. Viruses are pure nucleic acid with protein coats and tails; they cannot, however, reproduce themselves except in host cells—in this case, bacterial cells. E. L. Ellis and Max Delbrück had worked out techniques for using these viruses in genetic studies, and they showed that the viruses carried genes arranged very much like those in chromosomes in plants and animals. Alfred Hershey and Martha Chase at the Carnegie Institution in 1952 showed that a virus entered the bacterial cell in such a way that its protein coat and tail were left behind, attached to the cell surface, and only the DNA of the virus entered. They used radioactive isotopes to "tag" either the protein or the DNA and to trace the position of the isotope after the host cell was entered.

A pioneer in the study of viruses was the biochemist Wendell Stanley, who studied the disease called "tobacco mosaic" in which valuable tobacco leaves turned blotched and patchy. Stanley isolated the virus causing the disease by "crystallizing" it in pure form in 1935. He found that tobacco mosaic virus (TMV) was composed of ribonucleic acid (RNA) and proteins. By treating tobacco plants with the two substances he found that the RNA caused the disease but that the protein had no effect. This was the first actual proof offered that viruses were causes of disease, not just peculiar chemicals. Stanley founded the Virus Laboratory at Berkeley, California, where his work has been continued by Heinz Frankel-Conrat. Stanley won a Nobel Prize in 1946 for his pioneer studies.

While some viruses utilize RNA as the genetic complex, others apparently employ the related DNA, and, apparently, all cellular organisms use DNA. Perhaps this, too, is an evolutionary development.

The Genetic Code. Convinced that DNA was the active component of the chromosomes, James D. Watson and Francis H. C. Crick at Cambridge University, England, attempted to work out the structure of the DNA molecule. Other efforts along this line had been made, but no explanation could be offered that fitted the actions of the DNA.

In DNA and RNA, the two classes of nucleic acids, the building blocks are nucleotides. Each nucleotide is formed from a phosphate group plus a pentose sugar, and an organic base. DNA, deoxyribonucleic acid, has one less oxygen in its ribose sugar than does RNA, ribonucleic acid. The organic base of DNA is either a purine (adenine or guanine) or a pyrimidine (cytosine or thymine). These bases always pair up, with adenine joining thymine, and guanine joining cytosine; they are usually represented by letters *A, T, G,* and *C* for simplicity. In RNA, the pyrimidine uracil replaces thymine. (Recent work [1967] indicates that in some RNA, *A* can sometimes be linked to *G* when there are more purine than pyrimidine bases available.)

Each nucleotide unit is composed of some thirty atoms made up only of carbon, oxygen, hydrogen, nitrogen, and phosphorus. The unit is so small, less than one billionth of an inch, that it lies at the lowest limit of an electron microscope.

Watson and Crick, and Maurice Wilkins at King's College, London, simultaneously hit upon the idea that the DNA molecule is in a double helix (spiral), although their conclusions were based on different evidence. Watson and Crick were attempting to build a scale model on the basis of known chemical activity, while Wilkins used X-ray diffraction data; these men presented their papers in the British journal *Nature* at the same time in 1953, and they shared the Nobel Prize in 1962.

Structure of DNA. The picture given us can be compared to a number of familiar objects, such as a spiral staircase, or a ladder, or a twisted zipper. The nucleotide organic bases attach to their respective pairs by means of that special bond, the hydrogen bond, which is weaker than the usual chemical bonds. Adenine and thymine are linked by two hydrogen bonds while guanine is linked to cytosine by three hydrogen bonds. If these bonds are thought of as forming the stair treads, the organic bases are then attached to the pentose sugar connectors and the phosphate groups to form the stair rails.

An additional curiosity is that the sequence of nucleotides presents one strand that "reads" down in the same sequence as the other strand "reads" up; they are opposites in polarity. The twisting of the two strands around each other forms the configuration of a double helix, shown in Fig. 53. The strands of nucleotides may number from thousands, in viruses, to hundreds of millions in a single human cell.

Fig. 53. DNA Helix

If we are able only to link *C* with *G* and *A* with *T,* it might seem at first that the combinations offered would be rather limited. This is not, however, like trying to communicate with a four-letter alphabet; it is much more like signaling with Morse code, where there are only dots and dashes. It is the sequence of the dots and dashes that allows the telegrapher to signal any word in the English language.

It is astonishing that all of the DNA necessary to assemble a human being is present in the egg nucleus. If the threads were strung together, this "tape" of instructions would be about five feet long but less than one ten-millionth of an inch in diameter.

Replication. The Watson-Crick model suggests the way in which DNA may reproduce, or **replicate,** itself. The weaker hydrogen bonds tend to break more easily than the others. When the double helix begins to unwind, the hydrogen bonds separate just as though a zipper pull were running down its double track. Present in the cell are various nucleotide "spare parts." The hydrogen bonds of these new parts join the hydrogen bonds of the corresponding partners in the original double strands. The two original strands cannot return together; each strand is now paired with a new strand.

Consider once again the sequence of events in ordinary cell mitosis. The single strands that form in the new nucleus at the telophase more or less disappear from view during interphase. Each strand unwinds and, from the spare parts brought into the cell, a new partner strand is formed. As the cell enters prophase, the newly doubled strands become tightly coiled and are released from the nuclear membrane to move toward the equator of the cell. At metaphase, the double strands, lined up along the equator, "unzip." One half (one single strand) goes toward each pole as the cell continues mitosis. Exactly what controls the coiling and uncoiling is not known.

Duplication or replication of single strands of DNA has actually been induced in the laboratory by the addition of DNA to aqueous solutions of the bases, sugars, and phosphates needed. Arthur Kornberg and his associates at Washington University and Stanford Medical School have perfected this system, which won for Kornberg a Nobel Prize in 1959. Thus, we can observe how the cell duplicates itself exactly in this fashion, but other questions arise. How does the DNA pass on its instructions as to what work the cell is supposed to do?

The DNA remains in the nucleus and cannot go into the cytoplasm to "direct activities" when the cell is not reproducing. Therefore, a special kind of RNA, **messenger RNA,** does the job. One strand of DNA, or part of it, acts as a pattern or template and assembles an RNA strand that is like the opposite DNA segment. "Puffs" can be seen on the active sites of chromosomes in electron micrographs. These have been analyzed as being RNA. The messenger RNA molecule goes out into the cytoplasm to the ribosome structures that build proteins and acts as a template to form a specific protein, or its simpler chain, a polypeptide.

Let us say that the DNA active coding segment reads *C A T, C A T*. The opposite DNA segment would read *G T A, G T A* and the messenger RNA would line up in the same reverse order as the DNA: *G U A, G U A* (*U* because uracil replaces thymine in RNA). This is shown in Fig. 54.

Apparently, only one of the two strands of the DNA is used in coding information, the same strand each time. The messenger RNA goes out into the cell to the ribosomes and waits for amino acids to come along. Each amino acid has a place upon which to attach a code label segment, **transfer RNA,** three-unit nucleotides that have no partners. The code letters of transfer RNA are not definitely known, so letters used to represent their codes are hypothetical. The letters *ACC* and *G* are on the sites where the amino acids attach.

If the correct amino acids that have been picked up by the transfer RNA come along just then, the messenger RNA would promptly line the transfer RNAs together to form the polypeptide it is equipped to form. For example:

Coding DNA   — *C A T , C A T* —
   segment
hydrogen bonds — ... .
Opposite DNA   — *G T A , G T A* —
   segment
Messenger RNA — *G U A , G U A* —

Then the messenger RNA assembles the transfer RNAs attached to the amino acids:

Messenger RNA — *G   U A ,   G   U A*
               ... .
Transfer RNA   — *C   A U ,   C   A U*
amino acid   *AA*   *AA*

If the polypeptide consists of two units of the amino acid labeled with *CAU,* that chain is the one that is built, according to current thought.

Amino Acid Codes. There are some twenty primary amino acids that link together to compose the normal proteins of organisms. You recall that proteins are formed by the union of amino acids. The amino (NH_2) and carboxyl (COOH) groups link in a peptide bond COHN, squeezing out H_2O. Peptides are joined together into polypeptides, and polypeptides are linked to form proteins. The succession of carbons and nitrogens forms a kind of backbone, with hydrogen, oxygen, and the side groups (R) which hang from the backbone.

Linus Pauling and others came to the conclusion that these protein backbones were also coiled in a helix form, though more tightly coiled than DNA. Frederick Sanger of Cambridge University first analyzed the protein insulin as being composed of two protein chains, one sequence of twenty-one amino acid molecules, and the other of thirty amino acid molecules. Max Perutz and John Kendrew at Cambridge discovered that, in addition to the protein

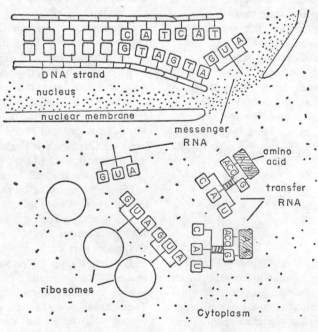

Fig. 54

being composed of peptide links which coil in a helix of carbons and nitrogens, the helix may also be folded in a complex manner. Each of these men received a Nobel Prize for his work in protein structure.

Because of the ways in which protein can be composed if all proportions and sequences of the twenty amino acids can be used, the possibilities for unique proteins are astronomical.

DNA has only a four-letter code and there are some twenty amino acids; the problem of finding out which DNA code stands for each specific amino acid has been intriguing. Since it appears that the DNA can be read in one direction only, three-letter combinations of the four letters *C, G, A,* and *T* could be united into exactly twenty different useful combinations. It must be emphasized that there is still much to be learned about this field, and some of the current ideas will no doubt be modified in the future. Four letters can be arranged in sixty-four different ways; since about twenty amino acids result, it is possible that more than one triplet can call forth a given amino acid, or that some combinations call forth nothing.

Severo Ochoa and his co-workers at New York University in 1955 discovered that artificial RNAs could be synthesized in a relatively simple manner by using substances from certain bacterial cells. Ochoa shared the 1959 Nobel Prize with Kornberg, who had been his student. Then, in 1961, Marshall Nirenberg and Heinrich Matthaei, at the National Institutes of Health, discovered that they could make a synthetic RNA composed entirely of uracil, *UUU.* They added *UUU* to a protein-synthesizing medium and succeeded in showing by radioactive tagging that the *UUU* picked up the amino acid phenylalanine.

Since then, many workers have been trying to find out which code letters would call forth a given amino acid. Since Nirenberg used only one repeated letter, *UUU,* he had no difficulties involving sequence. Sequence would be more complicated to determine since *CAT* could be assembled *CTA, ATC,* and *TAC,* for example. Code RNA words for twenty amino acids have now been tentatively proposed. There is some indication, however, that the first and third letters are more important than the middle letter. A number of polypeptides have now been assembled by mixing the messenger RNAs in different ways.

This research is unraveling some of the mysteries of the cell that seemed impenetrable only a few years ago. It is virtually impossible to give adequate credit to all the people who have contributed to current research; to name some, and not all, is probably unfair, but not to name any of our modern pioneers while revering such names as Aristotle, Leeuwenhoek, and Pasteur is even more unfair.

The information thus far uncovered has opened vistas revealing the possibility of controlling genes and gene defects. This may not be as simple as it appears, however, for there has been little success in coding single specific genes.

James Sumner of Cornell first isolated an enzyme in pure crystalline form in 1926 and showed that enzymes are proteins. It is now clear that all proteins, as well as enzymes, are synthesized in the cell by DNA and RNA. Segments of the RNA can be recognized which collect certain amino acids that can be linked together into polypeptides. Enzymes may be composed of more than a single protein chain, or a polypeptide, and it appears that different locations on the DNA are responsible for differing chains. The idea that one location or one gene was responsible for one enzyme has then been modified to indicate that one gene is responsible for one polypeptide chain.

A number of questions remain to be answered, in addition to breaking gene codes like the amino acid codes which are being solved. There is the problem of the influence of the cytoplasm on the process. Experiments have shown that the cytoplasm may reject "foreign" instructions, or its metabolic balances may be severely affected by them. Some evidence, as yet tentative, indicates that genetic factors are carried in the cytoplasm. While it is true that the cell cannot grow and reproduce without its nucleus, factors may be carried along in the cytoplasm at cell division which could affect the ability of the nucleus to carry out its direction, or could, in some other way, control cell function.

Differentiation. Another puzzle that has been unsolved is the manner in which differentiation is controlled in the cell. We know that the fertilized egg cell contains all the requisites for developing a complete organism. Experiments have demonstrated that during early cell divisions in some animals each cell is still capable of producing a complete organism. And in tissue culture, the cells continue to divide and to reproduce identical cells as long as nutrients are supplied. Yet, as cell division continues in the normal animal, the cells soon become quite limited in their abilities to take on different jobs and perform only specific functions.

For example, very early in the development of an embryo one area of the surface differentiates into a structure called the neural tube, which eventually forms the nervous system. This area is directly above a particular region of the developing gut on the interior of the animal. If a piece of this gut area is transplanted to another area beneath the surface, that surface area will also form a neural tube, even though normally it would only form body surface ectoderm. This ability lasts only a very short time, however. After this period, the surface no longer has the potential to do both jobs. Something has "committed" the tissues to their specific roles. All tissues must reach a stage during development when they are "committed" to be heart or lung or muscle, and cannot in general become anything else.

At the time of Weismann's germ plasm theory, it was thought that only the germ or reproductive cells carried the total message, and that somatic (body) cells carried only the portions they would need for their missions. We know now that this is untrue, for every cell has the same number of chromosomes and the same quantity of DNA regardless of whether it becomes kidney, heart, or muscle tissue. The quantity of RNA is greater in some germ cells, but the significance of this is not understood.

We have indicated the presence of proteins as well as the nucleic acid DNA in the chromosomes. These proteins, called **histones,** are unique and are not found elsewhere. Since 1950, it has been speculated that the histones in some way modified the activity of the DNA.

James Bonner and Ru-Chih Huang of the California Institute of Technology have developed evidence that histones perform their controlling influence by blocking certain parts in the DNA strand. Since bacterial cells are identical, it would be impossible to study differentiation with these organisms although they have been used in many experiments on the nucleic acids.

Bonner and Huang, using pea plants, extracted the DNA from pea seeds, but it always had the protein histones attached. This indicated that the DNA does not exist free in the multicellular organisms that show differentiation. When the DNA with histones attached was placed in a test tube with a mixture of sugars, phosphates, and organic bases, the pea-seed DNA made messenger RNA as expected. When the histones were stripped from the DNA, however, a much larger quantity of messenger RNA was made, indicating that the histones had prevented maximum RNA manufacture in the previous experiments.

The conclusion was that the histones act in such a way as to mask certain segments of the DNA, so the messenger RNA sent out to ribosomes in the cytoplasm will make only a certain protein. It appears that the ribosomes copy any message brought to them by the messenger RNA. Pea-seed DNA plus histone sends out the message to make a pea-seed protein that is not found anywhere else in the pea plant. In the same way pea-bud DNA plus histone makes only pea-bud protein. But if the histone is stripped from the pea-bud DNA, the pea-bud DNA sends messages to make pea-seed protein when the pea ribosomes are added to the culture.

Some evidence of the ability of ribosomes to copy whatever recipe is presented was developed by Bonner, Huang, and R. V. Gilden. Instead of using pea ribosomes this time, they extracted ribosomes from a common intestinal bacteria, *Escherichia coli.* The pea-seed DNA added to the culture medium that contained *E. coli* ribosomes turned out a protein strongly resembling pea-seed protein.

There is some indication that these various processes are essentially accomplished by what is known as a **chemistry of surfaces.** Just as the key-maker takes a pattern or template and cuts the blank metal according to the model, the messenger RNA may be assembled by modeling itself from the surface of the DNA strand. Then, just as the new key opens a lock by pushing a series of plates into place, the messenger RNA pushes the transfer RNA into place to assemble the amino acids. Apparently many chemical reactions occur because the surface configurations of one molecule somehow "fit" the surface configurations of another molecule. Enzymes also perform their functions because of similar surface patterns. In the cell nucleus, the surfaces of the DNA strands are copied by the messenger RNA. This copy is carried to the ribosomes. The transfer RNA copies the surface message presented by the messenger RNA and goes in search of an amino acid that carries a corresponding or matching message, as we have already seen.

The proposed activity of the histones in blocking or uncovering the segments of DNA adds a new puzzle to reproduction beyond finding the code words for the amino acids and genes. The mechanism by which the histones can shift their locations is entirely unknown.

Research suggests that some of the protein enzymes or hormones are exported from one cell or group of cells to the next, perhaps via the endoplasmic reticulum. These substances would affect the covering of the DNA by moving specific histone segments, which would alter the "recipe" that was carried to the ribosomes of that cell. In turn the second cell or cell group would export a different substance that would unmask a different histone portion in a third cell or group of cells. In this way, differentiation could be accomplished.

François Jacob, Jacques Monod, and André Lwoff of the Pasteur Institute of Paris were awarded the Nobel Prize in 1965 for research on how metabolism is controlled. They used micro-organisms for their experiments in order to get a single cell and rapid generations. They proposed a system model with three classes of genes: structural or synthesizer genes, operator genes, and regulator or suppressor genes. One gene functions as an enzyme-maker while operator genes are adjacent to it. The operator genes are seen as "turning on" the synthesizer gene. A regulator or suppressor gene would then inhibit or "shut off" both the synthesizer and operator genes at the proper time, thus controlling function.

This scheme is variously known as the "Operon Theory," the "sequential enzyme" system, and a "feedback" system, all descriptive to some extent. Investigations in laboratories around the world are demonstrating that systems in other organisms follow the model proposed.

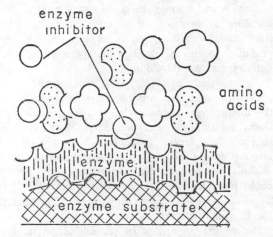

Fig. 55. Hypothetical Model of Function of Enzyme Inhibitors

Mention should be made of studies on enzyme inhibitors. When enzyme production is under way, a larger supply than necessary may be produced. Production of enzyme inhibitors would tend to regulate the reaction which is occurring, and to stop or slow it when the desired results have been produced. The organism does not function as an off-and-on manufacturing plant. It functions continuously, with interplay or feedback performing the necessary checks and balances. This steady state is vital to the organism. See Fig. 55.

Evidence suggests that histones participate to control the DNA, thus allowing for differences between the various kinds of cells. Conversely, failure of the histones to properly block the replication could be the cause of metabolic disorders. Cancer, for example, is characterized by a wild uncontrolled reproduction of cells, as though the brakes to stop runaway duplication had failed.

There are a number of places in the system where an error could lead to results differing from the instructions in duplication. Some of the congenital enzyme deficiency diseases show that perhaps only one amino acid differs from the normal, but it renders the enzyme useless. Suppose the messenger RNA copied *UUU* as *UUA* in just one set of instructions—the entire protein might be changed.

Ultraviolet radiation is readily absorbed by nucleic acids and apparently can cause a rearrangement of the nucleic acid molecules which would produce changes in the pattern for the cell. The changes that occur in genes are referred to as **mutations,** and have an important bearing on the variations produced in living organisms, as we shall see later.

The DNA structure helps to explain the change a virus can make in a cell. Some bacteria have their DNA arranged in a single ring "chromosome" structure. When a **phage** (a virus that attacks bacteria) penetrates the bacterial cell, the virus DNA breaks into the bacteria ring DNA and becomes part of it. It is then duplicated when the bacterial ring DNA untwists to replicate. The DNA is so altered that, for example, the ability of the bacteria to grow on its usual medium may disappear. Later the bacteria may eject the virus DNA and resume normal activity, or the virus may kill the bacteria.

The size of the virus DNA molecule is about the same as the DNA recently discovered in cytoplasmic organelles such as chloroplasts and mitochondria. This has increased speculation that the organelles may originally have been parasites that became permanent residents.

New research indicates that ribosome structure can be altered so that the ribosome will produce different proteins than those originally directed by the DNA. The drug Streptomycin apparently alters bacterial ribosomes. In this way drug-resistant strains may develop.

BISEXUALITY—A SUCCESSFUL ADAPTATION

It is quite possible for a simple organism to reproduce itself by dividing in half. We have seen the process called mitosis in which cells reproduce by the nuclear heritage of the cell dividing into two parts. This method, while quite successful for a single cell, is hardly adaptable for larger animals or plants; one can hardly imagine a direct division of a human being taking place.

Some cells are able to get more offspring from a single parent cell by repeatedly dividing the nucleus and then dividing all the adult cytoplasm among the daughter nuclei, destroying the parent cell. This is usually called **sporulation.**

Another possibility for reproducing is **budding,** in which an outgrowth begins on the parent plant or animal and grows into another adult, usually dropping off the parent when it becomes mature enough to survive alone. This is still a cell division by mitosis, however, and will contain exactly the same nuclear heritage in each cell as the parent had. Many of the lower plants and animals follow this method of reproduction part of the time.

The drawback to a one-parent inheritance is that the offspring are committed to exactly the same pattern of structure the parent has, unless some genetic accident occurs. Thus there is little possibility for variation in the line, and this markedly lessens the chances for survival of the organism if conditions in its environment should be changed. Variation in the organism is the only means by which a group can have a chance of surviving over a long time; through variation the offspring can adapt successfully to new conditions.

Originally, reproduction must have been asexual, by a simple division or the other related means: sporulation and budding. It must have been early in the existence of organisms that the mixing of genetic materials began to occur. Perhaps this began simply because of the ease with which some of the protein bonds can be broken in the long chains, causing them to reunite with different protein segments.

We know now that bacteria sometimes cling together and exchange part of their nuclear material through a cytoplasmic bridge. This has not been observed yet in the blue-green algae or in viruses, but some kind of exchange occurs in organisms above this level. Usually, these are even exchanges—straight transfers of material—but in many simple organisms one of the two abandons its "body" and the nuclear material joins that of the partner. In these ways, a recombination of the nuclear inheritance can be accomplished before the cell divides into two daughter cells. The real beginning of sexuality among plants and animals lies at this level. At first, the organisms that underwent recombination of nuclear material were probably all alike, but eventually differences began to arise that were advantageous in making sure that partners were found. We refer to the various mating types as **gametes;** when both gametes are alike they are **isogametes,** and when the two gametes differ, they are called **heterogametes.**

The change from isogametes to heterogametes was, no doubt, gradual. We can assume that among flagellate protistans there is no particular problem in finding mates because all are equipped with the means for locomotion—flagella. Yet the organisms are quite small, and there might be an advantage if half the population were immobile and the other half mobile so they could search for mates.

Assuming then that there was an advantage to having one less mobile partner in finding a mate, another advantage is readily apparent—the sessile partner could store more food to nourish the "new" organism if it did not have to scurry around. Thus, we find almost immediately advantages that can be seen to continue into higher animals as well as plants. All animals, and many plants, have male sperm which is flagellated, or motile, while the egg is relatively stationary and takes on extra food material.

There is no reason why there could not have been more than the two mating types, which we call male and female. In certain protists, the **Paramecium,** as many as eight different mating types have been identified. A variety of mating types probably occurred during the period in which heterogamy was developing.

Apparently **bisexuality,** or two mating types, proved to be all that was necessary in accomplishing the nuclear recombination needed to provide variations; certainly, it was the system that was most successful in surviving among all the living organisms. Perhaps too many mating types increased the difficulty in finding the correct mate. If Type *A* could mate with *B,* and *B* could mate with *C,* but Type *A* could not mate with *C,* it could hamper the locating of a correct mate considerably.

Before the beginnings of recombinations and exchange of nuclear material, organisms must have been equipped with only a single set of chromosomes. This condition is known as **haploidy,** meaning half the double set that investigators had considered to be normal as a result of their investigations among higher animals. The normal possession of double sets, or pairs, of chromosomes is known as the diploid condition or **diploidy.**

The change-over from the primitive haploid condition to the diploid condition in animals was gradual, but complete. Haploidy, however, has continued to be the normal condition for many plants for part of their life cycle. Some authorities feel that the existence of haploidy in an organism is more definitive than photosynthesis in identifying a protist as a plant, and that diploidy is indicative of animal stock.

Among some of the protists that have flagella, life cycles suggest the way in which diploidy may have arisen. The *Chlamydomonas* (Fig. 56) is a small ovoid organism bearing two flagella. For much of the life cycle of this flagellate, it is haploid and reproduces asexually simply by splitting lengthwise. Each daughter cell takes one flagellum and grows one new one. Then, when conditions change and winter approaches, two identical haploid adults (isogametes) merge into one cell with four flagella. Gradually, a cyst or case forms around this stage which is, for this time only, diploid. During the winter, however, the zygote, or fertilized cell, divides twice so that when spring comes four tiny haploid *Chlamydomonas* swim out. This variation of the life cycle undoubtedly arose as a protective device to avoid the rigors of winter. Initially the internal division during encystment may have been accidental: a lack of means of suppressing division during the long period. It is now firmly established as an advantageous pattern in the life cycle of these organisms. It is not known how the organisms "know" that winter is approaching, but they may be sensitive to the drop in temperature or to the amount of light as the days grow short.

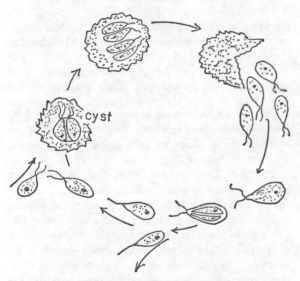

Fig. 56. *Chlamydomonas* Life Cycle

Although our study concerns animals, it is interesting to compare plants at a similar level of development with animals. The alga *Spirogyra* forms long strands of cells by asexual (vegetative) cell division. Occasionally, two strands will then move together and form cytoplasmic bridges between cells. The nucleus of one cell moves across the bridge and merges with the other nucleus to form a zygote in that strand. Instead of remaining encysted in the diploid phase, however, the nucleus immediately divides twice to form four nuclei. Three of the nuclei degenerate and the fourth one forms the haploid cell which will start a new strand.

At this primitive level we can see that the selection of a line of evolution for haploidy and diploidy was ready to be made. Many plants actually have two kinds of plant bodies in their life cycles, a growth of one type called the **gametophyte plant** during the haploid state, and a growth of another type, the **sporophyte plant,** which represents the diploid state. The haploid growths may actually have two kinds of growth as well if they are heterogametic.

Some plants underwent divergent evolution in the direction of having the haploid portion of the cycle predominate. For instance, there are haploid male moss plants and haploid female moss plants. Only when the male sperm cell swims over and enters the female cell structure does the moss enter a brief diploid phase. The fertilized female structure forms a spore cup and immediately divides the zygote into four cells that are thrown off as haploid spores to form new male and female gametophyte plants.

Most plants evolved so that the diploid phase became predominant, however, and the haploid phase became progressively reduced. Ferns, which are considered to have evolved into the modern seed plants, have diploid sporophyte bodies that are the dominant part of the growth. Yet ferns throw off haploid spores which grow into tiny gametophyte plants that bear the male and female cells. The male fertilizes the female in its cup and the diploid fern plant grows from the cup.

Animals did not continue to exhibit haploidy above the level of the protists. Diploid cells compose all multicellular animals. The reproductive cells alone are divided so that they will be haploid gametes until the zygote, or fertilized egg, is formed. We will discuss how this is accomplished by the process known as **meiosis.**

Modern seed plants, by convergent evolution, have developed a similar condition. The diploid plant bodies produce flowers which contain male and female haploid cells. The male cells (pollen) fertilize the female cells in the flower ovary to form a diploid zygote. The seed that grows in the fruit and, subsequently, in the ground is a diploid plant (sporophyte). No haploid growth beyond the actual male and female cells occurs.

MEIOSIS

Each animal parent has a diploid number of chromosomes in each body cell, including the so-called **germ cells**—the specialized cells which produce the cells that create the offspring of the parent.

The diploid condition means that all the chromosomes come in pairs; thus, there are two chromosomes that carry each characteristic. The paired chromosomes are not identical, however, because one was contributed by the father and one by the mother. For example, if your father has brown eyes and your mother has blue eyes, the chromosome pair carrying the gene for eye color will have one chromosome with your father's eye trait and one with your mother's eye trait. The chromosomes will be identical as far as their function is concerned: both will govern eye color. These are called **homologous chromosomes.** But each will be different as far as what eye color is specified. The two members of the pair of chromosomes are called **alleles,** or **allelomorphs,** meaning "other forms." Each chromosome is the allele of the other in a pair of homologous chromosomes. We can represent this by saying that E is the hypothetical chromosome for eye color. The allele from the female can be represented by E^t and the one from the male by E^m. The pair of chromosomes would be: E^tE^m.

In those animals that are bisexual a new individual is formed when the sperm, the male cell, merges with the egg, the female cell. This process is called **fertilization.** The fertilized egg is called a **zygote** to distinguish it from an unfertilized egg.

If the egg and the sperm cells each contained the diploid number of chromosomes, the fertilized egg would have too many chromosomes. In man, the diploid number is 46, or 23 pairs. If each parent contributed 46, the zygote would have 92 chromosomes, and the number of chromosomes would keep doubling with each generation. Soon the nucleus would be bursting with excess chromatin, its messages confusingly duplicated. Since we know that the number of chromosomes stays the same for a given species, a process must take place which solves this problem. **This process is called meiosis, a specialized kind of cell division that occurs only within the reproductive organs.** Each germ cell is reduced by division until it carries only half of each pair of chromosomes; that is, the haploid number of chromosomes which, in man, is 23. If n represents the number of chromosomes in a given species, the diploid condition is said to be $2n$, and haploidy is n. In man, diploidy would be 2×23 or 46.

Meiosis, shown in Fig. 57, has, essentially, the same phases as mitosis, ordinary cell division; however, there are a few important variations. The prophase stage of meiosis is extended over a period of time long enough to distinguish several separate stages that occur within it. During prophase, one very important difference in meiosis occurs.

In the prophase of ordinary cell mitosis, the alleles of a pair of chromosomes ignore each other when they are duplicating the strands. Nor do they react to each other when they migrate to the midplane of the spindle for metaphase; they line up at random. During prophase of meiosis, however, (called **first prophase)** the alleles are attracted to each other and join together at the kinetochores—the activity

centers of the chromosomes. Thus, the alleles, or homologous pairs, act together.

Preleptene. The first stage within meiotic prophase is called the preleptene (also known as proleptotena and preleptonema). The chromatids, single threads, condense and coil much as they do in mitotic prophase. Then they uncoil somewhat and stretch out into threads (chromonemata). Chromomeres, which are darker areas, like beads, show in stained preparation; these are now thought to be areas where the coils are tighter.

Leptotene. As prophase continues, the cell is in leptotene (leptonema) stage. The nucleus swells and the nucleolus doubles in size, indicating that DNA, RNA, and histone synthesis are going on actively. During this phase, the chromatids have been observed to form double strands in many organisms. It was formerly believed that doubling did not occur until later on, in metaphase, but it seems logical that the activities of the nucleus and nucleolus indicate synthesis during this stage even when duplication has not been directly observed. The doubled chromatids are called **dyads.**

The free ends of the chromatids appear to be attracted to the centrosome which lies outside the nucleus, so that the ends are clustered in the nuclear area nearest the centrosome. This is called a **bouquet formation;** it places the kinetochores close together so the alleles, or homologous chromosomes, find each other easily.

Zygotene. The zygotene (zygonema) stage follows as the doubled strands meet homologous chromosomes. The double strands form pairs, called **tetrads,** and the chromosome number is now $4n$. The actual pairing process is known as **synapsis.** (A synapse is a junction.) The kinetochores of the strands hold the pairs together by an undefined attraction. It appears that the distance between spirals in the coiled chromosomes, as well as the diameter, or **gyre,** of each loop, may change, making the chromosomes seem shorter or longer at different stages.

The chromosomes are active during these stages and the tetrad strands frequently move around and may cross over each other like a tangle of spaghetti, since they are held together only by the kinetochores.

Pachytene. The next stage, the pachytene (pachynema) is the most easily observed of the prophase stages in stained preparations. The chromosomes thicken (*pachy* means "thick") and stain lightly, with darker centers in the chromomeres. Crossovers (called **chiasmata**) can be seen.

Diplotene. The diplotene (diplonema) stage occurs when the kinetochores of the tetrads begin to repulse, rather than attract, each other. The strands stretch and would break apart from each other if

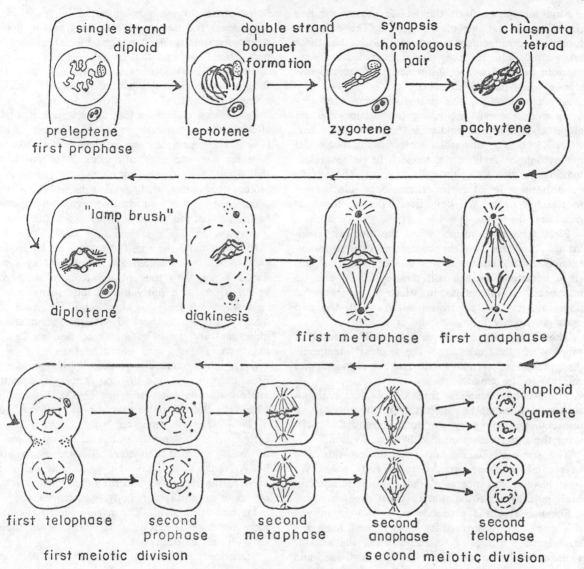

Fig. 57. Meiosis

it were not for the points of contact (chiasmata) where the strands have crossed over.

Some authorities believe that the strands actually break apart at the points of contact and re-fuse. This means that the alleles may no longer be "pure"; a hypothetical chromosome allele A^f may exchange part of its length with its allele A^m if the broken ends fuse with their opposite partners.

The repulsive force arising between the kinetochores during the diplotene stage continues, causing loops to bulge out between chiasmata, or causing the ends to flare apart beyond the contact points. The longer the chromosomes the more contact points there are.

In females, the diplotene stage is difficult to ob-

serve. The chromosomes are said to have a "lamp brush" configuration because they become fuzzy and look diffuse. This is apparently caused by extensive growth in the cytoplasm, which masks the chromosomes.

Diakinesis. The last prophase stage, diakinesis, occurs as the chromosomes shorten again and the tetrads migrate to the periphery of the nucleus. The nucleolus disappears and the chromosomes are darker-staining, as if they had picked up the remaining nucleolar material.

Outside the nucleus, the centrosome divides and the centrioles migrate to the poles. The nuclear membrane breaks down and the spindle is formed as it is in mitosis.

First metaphase finds the tetrads moving to line up along the equator of the spindle. The only difference in arrangement is that, in meiosis, the alleles are paired, whereas they are separated and lie at random in mitosis. This difference is of great significance when the cell divides.

At **first anaphase** the paired chromosomes separate and go toward opposite poles; our hypothetical allele A^t goes to one pole and A^m goes to the other. Both A^t and A^m are still double strands, but when the cytoplasm divides, the alleles will be separated, making the two daughter cells different. The tetrads have been reduced from chromosome number $4n$ to the two-strand, $2n$ condition. This is known as **reduction division.**

First telophase occurs when the nuclear membrane reappears and the cytoplasm divides in cytokinesis. The chromosomes do not uncoil at this stage because they are still double strands and do not need an interphase in which to prepare for doubling, as they do in mitosis. Instead, the two cells go directly into a **second prophase.**

The second prophase is very short. The centrioles migrate to the poles and the nuclear membrane disappears. The spindle forms and the chromosomes migrate to the equator.

The **second metaphase** finds the bivalent chromosomes $2n$ lined up at random along the equator. In one cell, we have only our double-stranded A^t; A^m is in the other cell produced by first division.

The strands separate into single chromatids, and in **second anaphase** proceed to opposite poles. We now have n chromosomes, the haploid number, at each pole by an equational or **equal division.**

Second telophase proceeds as it does in mitosis this time. The chromatids de-spiral and form the strands of the nuclear reticulum while the nuclear membrane also forms. The cytoplasm is divided and separates into haploid gamete cells.

It should be emphasized that the process of meiosis has variations in its pattern according to various species. For example, in some species of flies the alleles pair and form tetrads before they are visible in prophase. In some animals the tetrads tend to separate at the first metaphase so that the double strands become single strands and equal division occurs. Half of A^t and half of A^m go to each pole in this type of division. Then, at the second metaphase, the single-strand alleles separate so that reduction division takes place.

It has been calculated that the chances of all the alleles of either the male or the female going to one pole is $1:2^{23}$, or about one in seventeen million, not counting the exchange of pieces of chromosomes that occurs due to crossing over.

The production of haploid gametes by meiosis occurs only in the reproductive organs, the **gonads.** In the female, the gonad is an **ovary;** in the male, it is a **testis.**

In the female, the **egg** (or ovum) is the haploid gamete which is formed. The number of eggs produced varies with the species, from a single egg which is produced fairly infrequently to millions of eggs which are laid in a short time. The male produces **sperm** as the haploid gamete. Sperm are infinitesimal and must be motile to find an egg, so many more sperm are produced than eggs.

Even if only one egg is actually produced at a time, the same meiotic division occurs. The "extra" nuclear material that will not be used is cast off with very little cytoplasm at the time of each cell division. One daughter cell will get most of the cytoplasm at the first division, and the other will be a small discarded daughter cell nucleus, a **polar body.** At the second division, again, one cell keeps most of the cytoplasm and the other nucleus is cast off as a second polar body. Sometimes the first polar body will divide its nuclear material again to form the haploid second polar bodies even though they will all be discarded.

Statistically, then, each germ cell that undergoes meiosis produces two cells which further divide into four haploid cells. The female will produce one large egg cell and three small polar bodies; the male will produce four very small sperm cells, all of equal size. When one sperm cell penetrates one egg cell, the nuclei merge to form a single diploid cell—the zygote, or fertilized egg.

CHAPTER 6

HEREDITY

That life begets life is an observation which has been made many times and demonstrated in many ways through the years. It has also been seen that offspring are essentially like their parents, that like begets like. Yet the offspring, while essentially like the parents, may be in many minor ways different from the parents. It is to be expected that these observations would lead to endless speculation as to how such effects might be accomplished.

MENDEL'S CONTRIBUTION

Gregor Johann Mendel (1822–1884), an Austrian monk, was the first to work out an applicable theory which made it possible to predict the characteristics of the offspring when those of the parents were known. Mendel did not accomplish this work without effort, for he spent some ten years patiently planting various pea seeds in the monastery garden, observing their growth, and laboriously counting each plant, blossom, or seed that grew.

Mendel had been a student in Vienna when the controversy of evolution was spreading over Europe but he returned to the monastery at Brünn in 1851, well before Charles Darwin published his theories in 1859. Needless to say, Darwin's work had been preceded by years of discussion among scientists, for ideas rarely burst forth from nowhere but are a product of their times. We will discuss Darwin's work more extensively later.

Speculation on possible mechanisms for inheritance centered in Mendel's time on the theory of pangenesis. Some kind of particles or germules were thought to be sent from all the various parts of the body to the egg, there to be incorporated and passed on to the next generation. Thus, the offspring were blendings of their parents, a theory which appealed greatly to Darwin. It also allowed those who advocated the inheritance of acquired characteristics to push their views. Chief among these was Lamarck, of whom we shall also talk later. The foolishness of saying that a child could inherit strong muscles that were produced by the exercises of the father was recognized as early as the time of Aristotle, who remarked that it was obvious that if a man's arm were to be accidentally cut off, his child would nevertheless be born with all his limbs. Where would the particles to form the arm come from if the germules were sent from all parts of the body?

Whatever Mendel's intentions or ideas were at the beginning of his experiments, he picked his experimental plants well, for he was able to investigate a single set of contrasting traits at a time. He used seven different sets of single characteristics in all, and, each time, the results showed a similar percentage of like or contrasting offspring. Two contrasting traits of seeds were used: round smooth seeds versus wrinkled seeds; and yellow seeds versus green seeds. With pea flowers he compared red with white blossoms, and he also contrasted the locations of the blossoms; they occurred along the stem (axial) or clustered at the end (terminal). The pods also offered two contrasting sets of characteristics: they were either green or yellow, and they were either full or shrunken. He also compared long-stemmed plants with short-stemmed ones.

THE CONCEPT OF DOMINANCE

In all of Mendel's experiments, or crosses, he found that in the first generation of offspring (notated F_1, or first filial) all of the plants or seeds would be like one of the parents and none would resemble the other parent. One parent's characteristic was obviously stronger than, or **dominant** over, the other. In the second generation (notated F_2, or second filial), a few of the grandchildren showed the recurrence of the characteristics of the weaker or **recessive** parent. The F_2 offspring, however, showed the strong or dominant form in a ratio of 3 to 1 over the recessive form. In all, Mendel counted nearly 20,000 pea plants in order to be sure of tabulating sufficient numbers to show a valid pattern.

Although Mendel had no knowledge of chromosomes or genes, he came to the conclusion that each plant contained paired factors of some kind that controlled color or shape and that these were paired because the father had contributed one and the

mother had contributed the other. He concluded
that these paired factors (now called genes) separate
or segregate somehow when the germ cells form so
that each germ cell receives only one factor.

Mendel did not know about meiosis, discussed in
the previous chapter, but he realized that some sort
of separating process must occur.

A simple square, called a **Punnett square,** is used
to diagram the results when two adults with a single
opposite trait are mated. Let us use the symbol T
for the tall plants and t for the short plants. Assume
that we have picked pure-breeding adults that we
know never produced any variants and so would
bear two identical factors from each of their parents. Thus we have TT mated to tt in the parent or
P generation: $TT \times tt$.

The gametes will separate during meiosis so we
may have the possible gametes T or t from each
parent.

	T	T
t	Tt	Tt
t	Tt	Tt

Now if we fill in the squares by working down and
across, we have the gametes in the combinations
that can occur when a sperm fertilizes an ovum.
All of the F_1 offspring of this P_1 cross are identical.
Since all the plants are tall, we also see that tallness
is a **dominant trait,** but they have each carried along
a **recessive trait,** so we say that they are **hybrids**
or **heterozygous.** A zygote, you recall, is a fertilized
egg which carries a diploid set of characteristics.

Next, we take two of the F_1 hybrid plants, Tt,
and cross them. Again, we can plot the separation
of the gametes in the square:

	T	t	
T	TT	Tt	3 tall: 1 short
t	Tt	tt	or 1 TT: 2 Tt: 1 tt

We have one pure tall offspring (TT); two tall, but
hybrid, offspring (Tt); and one pure short offspring
(tt) in our "grandchildren"—the F_2 generation. As
far as appearance is concerned, we have a ratio of
three tall to one short offspring. The pure offspring
is called homozygous and the hybrid offspring
heterozygous.

We can show further that the hybrids carry both
the dominant and recessive traits by mating one
hybrid (Tt) with a pure tall (TT) in what we call a
back cross or **test cross:**

	T	T	
T	TT	TT	2 TT: 2 Tt
t	Tt	Tt	

All of the offspring will be tall. Or we can cross a
hybrid (Tt) with a pure short (tt):

	T	t	
t	Tt	tt	2 Tt: 2 tt
t	Tt	tt	

Here we obtain two tall offspring and two short ones.

Mendel tried all possible combinations with pea
plants and his results were always consistent with
his expectations. It made no difference in the inheritance of the particular characteristics he used
whether the parent which carried them was the
male or the female. However, later researchers
showed that some characteristics are definitely influenced by the sex of the parent that carries the
trait.

Mendel next tried combining in each parent the
crossing of two traits to determine if they interfered
with or influenced each other. He crossed pea seeds
that were yellow and round with pea seeds that
were green and wrinkled. Yellow (Y) is dominant
over green (y) and round (R) is dominant over
wrinkled (r).

The F_1 generation would be determined thus:

	YR	YR
yr	Yy Rr	Yy Rr
yr	Yy Rr	Yr Rr

All offspring are hybrid and they are all yellow and
round—the two dominant forms of the characteristics of color and shape.

When we cross two of the hybrids from the F_1
generation we have a random distribution, or segregation, in the gametes which gives us these possibilities:

> YR—yellow and round
> Yr—yellow and wrinkled
> yR—green and round
> yr—green and wrinkled

Plotted on a larger square we see the results of this
dihybrid (two characters) cross:

	YR	Yr	yR	yr
YR	YY RR	YY Rr	Yy RR	Yy Rr
Yr	YY Rr	YY rr	Yy Rr	Yy rr
yR	Yy RR	Yy Rr	yy RR	yy Rr
yr	Yy Rr	Yy rr	yy Rr	yy rr

At first glance the results in the square appear to be quite varied, but not if we consider the principle of dominance. If one *Y* or one *R* appears in a square, that offspring carries the dominant appearance. Nine squares contain at least one *Y* and one *R,* so these offspring would all appear yellow and round. Three squares contain *Y* and *r*—or yellow and wrinkled—offspring; three squares contain *y* and *R*—green and round—offspring. Only one square is fully recessive—it contains only the single green and round offspring. The numerical expression of this independent assortment is 9 : 3 : 3 : 1.

The appearance of the offspring is known as its **phenotype.** The actual genetic makeup of the individual, with whatever characteristics it may carry in addition to its dominant characteristics, is called the **genotype.**

Furthermore, only one pure dominant and one pure recessive occurs in the sixteen squares. This would provide for only a small number of individuals representing extremes in genetic makeup and thus tends to keep the population within a normal, or average, range. We will examine this effect in more detail later.

The results of multiple genes may also be calculated mathematically. The genes involved must not be linked, but must show **independent assortment.** That is, the genes must be on separate chromosomes so that they will move independently of one another at meiosis.

When gametes are formed, chromosome pairs are separated so that each gamete receives one of a pair of alleles. The probability of inheritance can be expressed thus: Given *Aa,* we have 50 per cent probability of getting gamete *A,* and 50 per cent of getting gamete *a;* or ½ *A* and ½ *a* (*Aa*=½*A*+½*a*).

Now, let us cross two *Aa*'s, and plot the offspring on a square, shown in Fig. 58. Multiply the fractions—vertical times horizontal gametes:

One-fourth of the offspring will be *AA;* ¾ will be *Aa;* and ¼ will be *aa.* This is, of course, the 1:2:1 ratio encountered previously. If *A* is a dominant characteristic, the ratio will become 3:1.

Mendel showed that when two hybrids with separate characters are crossed, such as the dominant

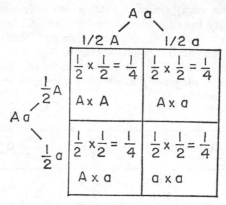

Fig. 58

yellow and round peas (*YR*) with the recessive green and wrinkled peas (*yr*), *YRYR*×*yryr* gives us the dihybrid *YyRr.*

By plotting the gametes into sixteen squares we saw that a ratio of 9 : 3 : 3 : 1 was developed. We can also calculate this arrangement mathematically by calculating each character separately. In other words, each character has a 50–50 chance of being present in the gametes. Suppose we have *AaBb* as our dihybrids to cross: *AaBb*×*AaBb*. The probability of *A* over *a* is ¾ to ¼ as shown in the single cross above. Likewise, the probability of *B* over *b* is ¾ to ¼. Now, assemble the characters and multiply the fractions across and diagonally, as Fig. 59 shows.

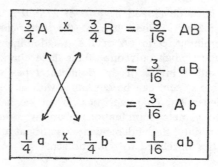

Fig. 59

Four-square and sixteen-square diagrams are not complicated. But if we try to cross three characters (trihybrids), we would need 64 squares in which to

$$\frac{3}{4}A \times \frac{3}{4}B \times \frac{3}{4}C = \frac{27}{64}ABC$$

$$\frac{1}{4}a \times \frac{1}{4}b \times \frac{1}{4}c = \frac{1}{64}abc$$

Fig. 60

plot the combinations. This is unwieldy and unnecessary because we can multiply as we did in the previous examples with less difficulty. Multiplying across we get the results shown in Fig. 60.

Other possible combinations seen above must be multiplied diagonally to obtain results such as those shown in Fig. 61.

$$\frac{3}{4} A \times \frac{3}{4} B \times \frac{3}{4} C = \frac{27}{64} ABC$$

$$\frac{3}{4} A \times \frac{3}{4} B \times \frac{1}{4} c = \frac{9}{64} ABc$$

$$\frac{3}{4} A \times \frac{1}{4} b \times \frac{3}{4} C = \frac{9}{64} AbC$$

$$\frac{1}{4} a \times \frac{3}{4} B \times \frac{3}{4} C = \frac{9}{64} aBC$$

$$\frac{3}{4} A \times \frac{1}{4} b \times \frac{1}{4} c = \frac{3}{64} Abc$$

$$\frac{1}{4} a \times \frac{3}{4} B \times \frac{1}{4} c = \frac{3}{64} aBc$$

$$\frac{1}{4} a \times \frac{1}{4} b \times \frac{3}{4} C = \frac{3}{64} abC$$

$$\frac{1}{4} a \times \frac{1}{4} b \times \frac{1}{4} c = \frac{1}{64} abc$$

Fig. 61

We have developed a ratio of 27 : 9 : 9 : 9 : 3 : 3 : 3 : 1. When two of the three characters are dominant, 9 in 64 of the two dominant types will occur; when only one of the three characters is dominant, 3 in 64 of the dominant types will occur. This system can be continued with any number of independent characteristics. When we listed the possible gamete combinations above we were considering only the independent assortment of characteristics unrelated to one another.

HEREDITY THEORIES

Unfortunately, Mendel's work went unnoticed when he published his results. The journal in which the findings were published had small circulation, but Mendel also corresponded with the well-known botanist, Karl von Nägeli, who was interested in heredity. Von Nägeli apparently did not perceive the significance of Mendel's results as an explanation for the variations required as possible mechanisms for change or for evolution.

Mendel discontinued his studies and reputedly became embroiled in local church squabbles. Although he lived after Darwin's work had been published and had been widely discussed in Europe, he never entered the scientific community again. Fortunately, his papers had been included in the listings of scientific literature where Hugo De Vries located them in 1900.

De Vries had developed the pangenesis theory, but he and others were searching for some type of variation that would be evident as a separate or discontinuous occurrence rather than an accumulation of extremely minute variations. These scientists had developed some evidence similar to Mendel's, but Mendel's explanation of the phenomena of inherited characteristics made their data intelligible.

Karl Correns in Germany and Erich von Tschermak in Austria rediscovered Mendel's work about the same time as De Vries did in Holland—each discovery was independent.

At the beginning of this century, many workers were investigating the various cellular activities of mitosis, meiosis, and the merging of the male and female nuclei to form the fertilized egg, the zygote. The search for minute variation in a series or progression which could change one species into another had not been successful. Researchers then began to consider the activities which were observed taking place among the chromosomes as possible mechanisms of heredity. Gradually, experiments concerning minute variations were abandoned as workers embraced the idea of large jumps (called **saltations** or **mutations**) to explain the origin of new species.

We saw earlier that August Weismann developed the germ plasm theory which held that the characters of a given species would be carried from generation to generation by means of special germ cells in each body. The other body cells existed only as carriers for the perpetuation of the species. He predicted that some kind of division occurred to reduce the total number of chromosomes to the haploid species number at fertilization. This process of meiosis was subsequently seen by the German zoologist Theodor Boveri, but Weismann's failing eyesight prevented him from observing it. Weismann felt that this reduction of chromosomes, followed by the fusion of the male and the female nuclei, would explain the mutations or variations that occur in inheritance.

One of the objections to Weismann's germ plasm theory was that the germ cells would have to carry an incredibly complex array of information. However, due to recent advances in knowledge of the cell and the structure of DNA and RNA, these objections are easily overcome. Previously, it was thought that the germ cell would have to contain all the information for all the individuals that had existed in the past plus all those which would exist in the future. Now we know that we need a code only for the animal type. The combination of the codes of the two parents could account for the individual differences occurring in each generation.

The insight that some scientists exhibited in the past is impressive. In many cases their ideas were discarded as unworkable, but new discoveries lure old ideas from oblivion and cast them in entirely new roles.

VARIATION

Francis Galton, Charles Darwin's cousin, tried to find the secrets of variation by means of statistics, but he missed Mendel's ratio completely. He did, however, initiate the science of biological measurements, **biometrics,** and the science of **eugenics**—the improvement of the race by selective breeding.

At the time that De Vries uncovered Mendel's work, Boveri and others, such as Oskar Hertwig, were investigating the functions of the sperm and the egg. Boveri worked out the genetic continuity of chromosomes, concluding that chromosomes persist in some form during the resting stage when they cannot be seen in the nucleus, so that the same number and form of chromosomes come forth for each subsequent cell division.

W. S. Sutton supported the ideas of T. H. Montgomery (both were American cytologists) in concluding that the paired chromosomes found in the zygote received one single-strand chromatid of each chromosome pair from the mother and one from the father. Each strand doubles prior to cell division, and every cell in the body then has the same kinds of pairs—half contributed by each parent. We could represent a hypothetical chromosome pair as: $A^fA^f+A^mA^m$. Here, A^f, received from the female parent, is a double strand; and its allele, A^m, received from the male parent, is a double strand.

Sutton reasoned that at meiosis, when the haploid gametes are to form, there is a random distribution of the alleles and that this accounted for the variations which occur. In other words, when the double strands form into pairs, or tetrads, the side on which they assemble is not predetermined, but occurs at random for each pair. Therefore, if we have two hypothetical chromosomes, A and B, they could be assembled at the equatorial plate in metaphase as shown below:

$$\frac{A^fA^f \qquad B^mB^m}{A^mA^m \qquad B^fB^f} +$$

When the cell division occurs, A^fA^f and B^mB^m go

to one cell, and A^mA^m and B^fB^f go to the other. The second meiotic division will separate the chromatids at random again, producing A^fB^m in two gametes and A^mB^f in two gametes. Here we find the maternal trait of one chromosome in the same cell with the paternal trait of another chromosome. In addition, the tangling and crossing over at the tetrad stage may mean the exchange of individual genes or groups of genes. Strand A^f may then have a small portion of A^m on one end while strand A^m would have the comparable A^f portion at its end, an arrangement such as shown in Fig. 62.

 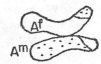

Fig. 62

Sutton saw that random distribution could account for the variations found in offspring who frequently resemble one parent in some factors and the other parent in others. This is not blood-mixing, but rather a sorting mechanism resulting in more variation. Sutton also realized that more than one factor was probably carried on a single chromosome; consequently, all these characteristics would be inherited together. These are now known as **linked characters** or **linkage groups,** and we will examine them in more detail later.

Most inherited characteristics are not due to a single factor, but rather to the sum of a number of factors. Mendel was either fortunate or wise when he chose the pea plant for his experiments, because it has a selection of single-factor characteristics that breed independently. Had he accidentally picked characters caused by multiple factors, or even characters that were in linkage groups, his work would have been far less significant and rewarding.

Mendel's work can be summarized by several "laws" which he put forth. The **law of independent assortment** indicated that, as in the case of the green wrinkled peas, the two traits are independent of one another. Each trait is a separate entity and acts apart from another so that the offspring are the sum of many traits sorted independently. We now know that this is true of some characters, but not of others. Independent action requires that the particular traits be governed by genes on separate chromosomes.

Mendel also observed the phenomenon of **dominance.** He saw that one allele can be dominant over another, masking the existence of the nondominant characteristic. We now know that this is also true only for some characteristics. For example, if Mendel had crossed red snapdragons with white, instead of using peas, he would have found his F_1 generation to be all pink instead of all red. The F_2 generation would then be 1 red: 2 pink: 1 white. Here neither red nor white is dominant over the other and so a blending is produced.

Another of Mendel's "laws" relates also to an example such as the one above. He referred to it as the **law of segregation.** At Mendel's time it was thought that offspring were a blood-blending of the parents. He stated that characters were not altered when they were mixed with others, but survived intact even though they might be masked by dominant factors. We know now that genes can be influenced by interactions but this does not make Mendel's "law" untrue, because the characters retain their integrity.

MODERN GENETICS

The union of Mendel's findings with later information on chromosomal behavior ushered in a period sometimes known as the era of the "new genetics." Morgan at Columbia University and his colleagues began studying genetics using fruit flies because they were easy to raise and had only four pairs of readily distinguishable chromosomes. During the next thirty years, a tremendous body of experimental data was assembled in genetics, based largely on studies with the fruit fly—*Drosophila melanogaster.* In 1933 T. S. Painter pointed out that a number of insects had salivary glands that contained cells with unusually large chromosomes. This advanced the study of genetics even more.

Sex Determination. One very important area concerns the chromosomal determination of sex. Early in the history of genetics it was found that all the chromosome pairs are composed of identical partners except for one pair. The pair that determines the sex of the animal has unlike mates in one sex of the animal but like mates in the other.

In *Drosophila* the sex chromosomes are called X and Y partly because of their shape. The females always produce paired X chromosomes, or XX; the male carries the unlike pair, or XY. How does this difference account for the ratio of males in a population? Using the square diagram again (Fig. 63) we can plot the gametes of the mother and father. (The symbol ♀ indicates female; ♂ indicates male.) Multiply both down and across.

We see that the female supplies two X haploid gametes while the male supplies one X and one Y gamete. The union of the gametes provides 50 per cent XX zygotes and 50 per cent XY zygotes, closely

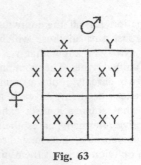

Fig. 63

approaching the natural birth ratio of females to males. The female, in cases like *Drosophila*, is described as homogametic, while the male is heterogametic. It also can be seen that the type of sperm which fertilizes the egg will determine at that instant whether the offspring will be male or female.

Man is thought to have a similar chromosomal situation. There are 23 pairs of genes in all, 22 of which are identical autosomes, plus one pair which carries the sex factor. These differentiated genes will be heterogametic if male, homogametic if female.

Other species of animals achieve sexual differentiation in varied ways. In *Drosophila*, if the female is XX and the male XY, we may assume that XX is required for femaleness. The lack of XX in the male might then be the determining factor for sex, rather than the actual presence of the Y in the male. The grasshopper, for example, has no Y chromosome. The female has the XX chromosome condition, but the male has only a single X, usually illustrated as XO for clarity. This means that if the female has 24 chromosomes the male has 23. By plotting the gametes (Fig. 64) we see that in this case XX produces the female and single X, the male.

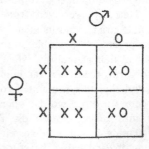

Fig. 64

In certain other animals, the female is heterogametic and would be represented by XY while the male is XX. Some authorities prefer to use different letters for this condition such as WW for maleness and WZ or WO for femaleness, in order to emphasize this difference in heterogamy.

There is currently some debate as to whether the *XX-XY* condition that has long been attributed to man is actually valid. Further research may change existing concepts, and science must always keep open the opportunity to do so. Only by remaining flexible can progress be maintained.

Another interesting means for sex determination is seen in some insects, particularly in the honeybee. This specialization involves **parthenogenesis,** or the ability of the egg to develop without being fertilized. In honeybees unfertilized eggs develop into males while the fertilized eggs develop into females. The females may be sterile workers or fertile queens, depending on the nutrition given them when young. Thus, the females have the diploid number of chromosomes, but the adult males are haploid. In the male, meiosis follows the usual steps, but the single haploid alleles cannot pair and separate. Nevertheless, division occurs according to the basic principle, with some of the new cells completely lacking nuclear material. These cells soon die and are resorbed. Since honeybees show a high degree of specializa-

tion in their social and physical patterns, it is interesting that they have still inherited the basic meiotic patterns, long since modified in the males but orthodox in the females.

Differences in ratios will be seen, depending on whether a trait is carried by the *X* chromosome of a male or a female. In *Drosophila* for example, we may cross a pure white-eyed female with a red-eyed male. Or we may cross a red-eyed female with a white-eyed male. The results are shown in Fig. 65.

Nondisjunction. One of the most interesting facets of recent chromosome research concerns genetic accidents in which a zygote receives more chromosomes than it should. The American geneticist Calvin Bridges did extensive research in this area which concerns the **nondisjunction of chromosomes.** Some types of mental deficiency and some diseases due to malfunction or failures in metabolism have been attributed to such aberrations. Nondisjunction occurs when a chromosome fails to attach to the spindle at metaphase and does not proceed to the proper pole. Then, when division of the cytoplasm occurs, the

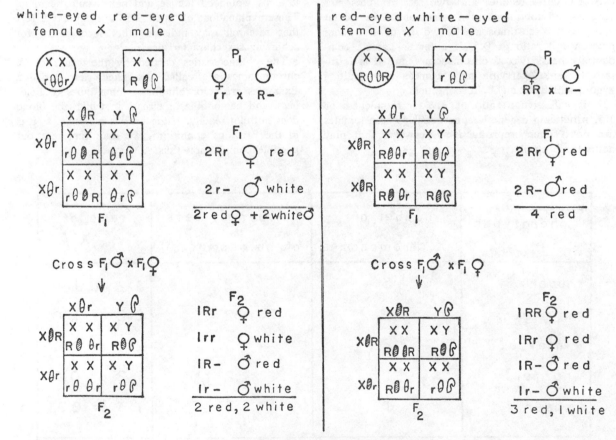

Fig. 65. *Left*–White-eyed Female×Red-eyed Male
Right–Red-eyed Female×White-eyed Male

wandering chromosome is incorporated into one or the other daughter cell. Sometimes such an accident will be lethal to the gametes; in other cases, the gametes will continue to function.

Mongolian idiocy, so named because eye folds similar to those of Orientals are present along with severe mental deficiency, is caused by the presence of an extra chromosome. This condition seems to be more frequent when the mother is past the age of forty, and is independent of the age of the father. Apparently, the chromosomes fail to separate normally during the long period of maturation of the oöcyte, and the infant carries 47 instead of the normal 46 chromosomes.

It is not known how common chromosome errors are in the general population; new techniques for examining cells are being introduced for use in diagnostic examinations in hospitals. Perhaps other defects or deficiencies will be uncovered that are also the result of abnormal chromosome numbers.

In *Drosophila,* nondisjunction has been observed in the sex chromosomes. Some authorities feel that sexuality is determined by the ratio of X chromosomes to the sets of autosomes. Thus in females two XX chromosomes and two sets of autosomes (the diploid condition) give a ratio of 2:2 or 1.0. In males, one X chromosome to two sets of autosomes gives a 1:2 ratio of 0.5. If a female gamete accidentally bears two X chromosomes, and it is fertilized by an X-carrying male gamete, the resulting zygote will have an XXX condition.

This will give a ratio of 3:2 for femaleness or 1.5, producing a genetic superfemale. These females are not "super" reproductively, however, but may actually be sterile.

Occasionally an entire set of autosomes will fail to separate properly, producing, for example, a ratio of two X chromosomes to three sets of autosomes; 2:3 or 0.66. This ratio lies between the 1.0 for normal femaleness and 0.5 for normal maleness, producing an intersex state. If only one X chromosome is present with the abnormal number of autosomes the ratio would become 1:3 or 0.33, below normal maleness of 0.5. This produces supermales, with accentuated male characteristics but reduced fertility. These conditions are summarized in Fig. 66.

Many genes occur on the X chromosomes and will obviously be affected by abnormal variation in the number of these chromosomes. It is thought by some authorities that the Y chromosome does not contribute significantly, perhaps carrying only factors influencing secondary male characteristics, such as amount and distribution of hair, and tone of voice.

If, in the first mitotic division of the female zygote (first cleavage), one of the X alleles should become lost, one cell would be female in character with XX, but the other cell would contain male characteristics of the XO condition. As the animal develops, half its cells would be female and half would be male. These explanations of "maleness" are unproved; further research may indicate other mechanisms for achieving sex characteristics.

The characteristics carried by the genes of X chromosomes are called **sex-linked characteristics.** Examples are color blindness, some types of baldness, and hemophilia, a disease in which the blood does not clot readily. If the gene is recessive, as it is in the previous conditions, we may represent our possible combination thus:

phenotypes	number of X chromosomes	number of sets of autosomes (A)	ratio of X:A
super	3	2	3:2 or 1.5
normal	2	2	2:2 or 1.0
intersex	2	3	2:3 or 0.66
normal	1	2	1:2 or 0.5
super	1	3	1:3 or 0.33

Fig. 66

1. Normal female (XX) mated to color-blind male (xY): See Fig. 67.

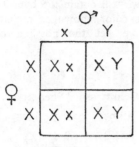

Fig. 67

Offspring: 2 normal males (XY) and 2 females carrying the recessive x gene (Xx) for color blindness. The normal X gene is dominant so both females are normal, but they act as carriers.

2. Carrier female (Xx) mated to normal male (XY): See Fig. 68.

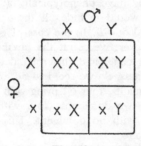

Fig. 68

Offspring: 1 normal female (XX), 1 normal male (XY), 1 carrier female (Xx), and 1 color-blind male (xY).

Recessive genes are relatively rare in the population. For example, color blindness afflicts between 4 and 10 per cent of male humans but less than 1 per cent of female humans. In order to produce female color blindness both parents must contribute a recessive gene $(Xx) \times (xY)$. See Fig. 69. If both parents are color-blind all the offspring will be color-blind $(xx) \times (xY)$. See Fig. 70.

Hemophilia functions in the same manner. The recessive gene is overcome in the Xx condition of the female, but appears in the male (xY) where there is no normal X chromosome to modify the recessive x. There is only one case on record of a female suffering from genetic hemophilia. Apparently the xx condition in hemophilia is lethal to gamete, zygote, or embryo, illustrating the way in which defectives are eliminated from the population.

Back Crosses. By examining the ratios of the offspring Mendel determined that chromosomes seg-

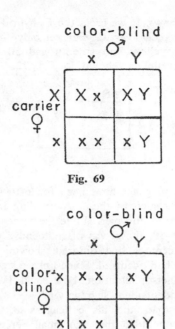

Fig. 69

Fig. 70

regated independently. Other investigators later discovered that not all characteristics separated independently—some remained linked and gave very different ratios. The linkage for any set of characters can be determined by performing test crosses, or back crosses.

Let us suppose we have a type of corn which shows a homozygous dominant form of yellow and smooth grains $(YYSS)$. The homozygous recessive is white and shrunken $(yyss)$. Crossed, the F_1 hybrid would become $YySs$.

If the characters Y and S are independent of each other (borne on separate chromosomes), the following gametes will result from the hybrid: YS, Ys, yS, and ys. Let us then perform a test cross, mating an F_1 hybrid to a homozygous recessive, shown in Fig. 71.

	YS	Ys	yS	ys
ys	Yy Ss	Yy ss	yy Ss	yy ss

Fig. 71

We would obtain one each of yellow and smooth, yellow and shrunken, white and smooth, and white and shrunken, or 25 per cent of each.

If the characters are linked (borne on the same chromosome), they will be unable to separate at meiosis, and the gametes from the F_1 hybrid $(YySs)$ can be only *YS* or *ys*. If we cross the F_1 hybrid with the homozygous recessive, we can get only the results 50 per cent yellow and smooth and 50 per cent white and shrunken, shown in Fig. 72.

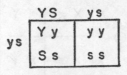

Fig. 72

A number of genes have been found to be linked by using tests such as these. A puzzling factor entered the picture however. While the percentages approached 50–50, there was often found a small remaining percentage that differed, showing that recombination had occurred. Since linkage would seem to be an either-or situation, how could 1 or 2 per cent show independent segregation?

T. H. Morgan and his colleagues at Columbia University concluded that the small percentage of recombinations that were observed represented crossing over during meiosis. We saw previously the tangling that takes place during the prophase tetrad formation (Fig. 62). If the crossover occurs between the genes *Y* and *S* in a small percentage of the matings, *Y* could then recombine with *s,* and *y* could combine with *S,* as Fig. 73 shows.

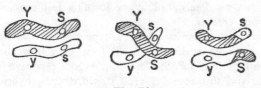

Fig. 73

Morgan also concluded that the frequency with which this would occur would be proportional to the distance between the two genes. In other words, the farther *Y* is from *S* on the same chromosome, the more likely a crossing-over would occur that will separate *Y* from *S*.

This principle of crossing over has been one of the most dynamic in modern genetics because it has enabled workers to map the locations of various genes on the chromosomes.

Polygenic Inheritance. Let us now consider another type of gene action in which the genes have cumulative effects—polygenic inheritance.

There are a number of characteristics, such as the size of fruit or the number of eggs laid by a hen, that depend on more than one gene. The total number of dominant factors of all the genes involved determines the degree to which the offspring will exhibit the characteristics. Skin color in man is one of

these characteristics, and has been found to depend upon at least two different genes.

If we cross a pure Negro (*AABB*) with a pure white (*aabb*), we will obtain the hybrid type *AaBb*, an intermediate color known as mulatto. Crossing *AaBb* × *AaBb* we get the possible combinations shown in Table II.

Table II

Dominants	Genotype	Phenotype
4	AABB	1 Pure Negro
3	AaBB ⎫	⎰ 2 Dark
3	AABb ⎭	⎱ 2 Dark
2	AaBb ⎫	⎰ 4 Mulatto
2	AAbb ⎬	⎨ 1 Mulatto
2	aaBB ⎭	⎱ 1 Mulatto
1	Aabb ⎫	⎰ 2 Light
1	aaBb ⎭	⎱ 2 Light
0	aabb	1 White

If these genes were independent of each other we know we would have phenotypically a 9:3:3:1 ratio. The nine would include all those that had at least one *A* and one *B*. In the case of polygenic inheritance, however, we count the number of dominants of either gene, and this total determines the degree of darkness. So we see, instead, one offspring with four dominants, four offspring with three dominants, six with two dominants, four with one dominant, and one with no dominant. This produces a 1:4:6:4:1 ratio.

If a ratio of this type is plotted on a graph we will see a bell-shaped curve, called the curve of nor-

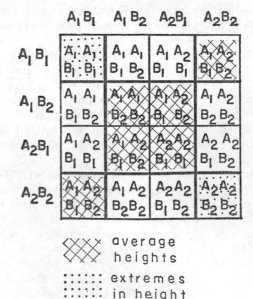

Fig. 74. Cross $A_1A_1B_1B_1 \times A_2A_2B_2B_2$

mal distribution. Variations of height, weight, and other characteristics of a polygenic nature also show this type of curve. See Fig. 83.

A similar situation occurs in the case of tallness, where two genes, neither dominant over the other, control height by intergradation. Instead of using large and small letters to identify the genes we will represent tallness by $A_1A_1B_1B_1$ and shortness by $A_2A_2B_2B_2$. In such a representation then, the more A_1's and B_1's present, the taller the individual. The manner of diagramming this cross is not important, as long as it expresses the cumulative effect of the genes. One method is shown in Fig. 74.

Prime factors:

4	$A_1A_1B_1B_1$	1	Tall
3	$A_1A_1B_1B_2$ } $A_1A_2B_1B_1$ }	4	Medium Tall
2	$A_1A_1B_2B_2$ } $A_1A_2B_1B_2$ }	6	Medium
1	$A_1A_2B_2B_2$ } $A_2A_2B_1B_2$ }	4	Medium Short
0	$A_2A_2B_2B_2$	1	Short

When chromosomes cross over, the exchange of alleles is not always a uniformly even exchange of genes. Sometimes a chromosome will actually be pulled apart; when it fuses again the order of the genes may not be the same or a segment may be omitted (**deletion**). If genes are missing, the gametes usually die. If inversion occurs, the chromosomes may have trouble pairing for synapsis. Occasionally, chromosomes will synapse with the wrong partners and, if crossing over occurs, neither chromosome will be able to pair up with its correct allele. Such events, known as **translocations**, are usually lethal.

Mutation. All of these activities are called **mutations** and give rise to changes in the phenotype and genotype of the organism, assuming the effects are not lethal. Mutations apparently occur in the normal course of reproduction, but they may also be induced in the laboratory by physical or chemical means. Chloroform, ether, caffeine, nicotine, LSD, and some hormones, vitamins, and antibiotics cause abnormalities when introduced into cells. Whether these substances actually reach the reproductive organs, or how much effect they might have on them, has not been determined.

It has been known for nearly forty years that X rays affect body tissues. Genes or chromosomes may be altered into a new stable configuration or mutation, or their capacity may be damaged irreparably so that mitosis does not proceed properly.

There is a considerable amount of natural radiation in the present environment. Primitive organisms, and perhaps early man, might have been subjected to more radiation because there was not as thick a layer of oxygen or as much carbon dioxide and other substances in the primordial atmosphere. In any case, it is estimated that about one in five humans carries some gene that has mutated due to natural radiation. This constitutes about one-tenth of the spontaneous mutations that are known to occur. It has also been estimated that the total of all X rays, radioactive substances, and fallout added to the natural level of radiation may eventually raise the incidence of mutation from the one in every five humans due to natural causes to two in every five.

Since there is considerable controversy over the effects of radiation it is strongly recommended that the reproductive organs be protected when X-ray pictures are taken for diagnostic purposes. Most hospitals now provide lead shields or aprons for patients and employees who are so exposed.

Some authorities hold that all abnormally caused mutations have a bad effect. Others believe that mutations are random in nature and natural selection will, as always, result in the beneficial mutations surviving and the poor ones being eliminated. The mutation rate may be only speeded up, and may not necessarily be harmful.

Recent evidence indicates that gene action may be influenced by such external physical conditions as heat and cold. The extent to which external forces may act on genes is undetermined. Experiments with plants indicate that one appearance may occur at a given temperature and an entirely different appearance at another temperature. How many genes may have their effects altered in this way has not been established, but the evidence indicates a wider range of adaptivity than was previously supposed possible without the alteration of gene structure or content.

Enzyme Action. Over fifty years ago Hans Driesch developed the hypothesis that genes produce enzymes, but he was unable to illustrate it. In the last twenty years it has been demonstrated conclusively that genes act by producing enzymes. Radiation of a given gene can stop its enzyme production and thus prevent the normal action of that gene.

All enzymes are known to be protein and proteins are characterized by their folded, complex structure and surfaces. The enzyme surface form determines the action it will have on other substances it encounters. If their surfaces interlock, the enzyme will aid the other substances in linking; if the surfaces do not interlock, no reaction will occur. This was diagrammed in Fig. 55. The enzyme itself, of course, is never permanently altered by the chemical reaction it catalyzes.

When the enzyme theory of gene action came under consideration, it was applied to the patterns of Mendelian heredity. At first the theory seemed to explain dominance as the presence of an enzyme and recessiveness as the absence of the enzyme. There were, however, many cases of incomplete dominance to consider. You recall that the red snapdragons, when crossed with the white, gave an F_1 generation of all pink flowers. A similar case is seen among Andalusian chickens, where a black fowl crossed with a spotted white one gives an F_1 generation hybrid called Andalusian blue. The F_2 ratio is 1 : 2 : 1 in both cases—or one dominant, two intermediate, and one recessive.

This indicates that enzymes must be produced by both the dominant and the recessive genes. The hybrid shows what seems to be a blending in appearance (the phenotype). However, the fact that the F_2 generation contains the pure recessive characteristic again indicates that the gene and its enzyme have not been changed or "blended" by mixing.

One of the characteristic functions of cells is the secretion of various protein substances which are unique to the species producing them. When a foreign protein, an **antigen,** is introduced into an animal, the animal will react by producing a substance, called an **antibody,** to eliminate the foreign protein. This reaction is apparently another example of the chemistry of interacting surfaces, as was the case with enzymes. The enzyme acts by helping to link two proteins together and then "unhooks" from both of them. The antibody links directly to the invading foreign antigen, rendering its surface inactive. It may cause a clumping, or coagulation of particles, a process called **agglutination.** See Fig. 75.

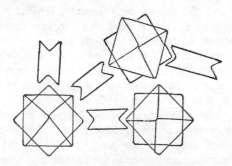

antigen + antibody → clump

Fig. 75. Antigen-Antibody Reaction

The antigen-antibody reaction is a major defense mechanism in the body. Antibodies are first produced by action of the thymus gland in the neck of the embryo; later they are produced by all the tissues in man, but especially by the spleen, the lymph nodes, and the liver. In the beginning stages of invasion of the human body by foreign organisms, antibody production increases slowly. It reaches a high level as the invaders are overcome more rapidly than they can reproduce, and the balance of the human system is re-established.

Often when antibodies for a given invader have been developed at a certain time, antibody production continues long after the invasion or disease is past. This defense surplus of antibodies is known as **immunity.** Some immunities are life-long; others last for a limited time.

Blood Types. We have introduced the antigen-antibody reaction here because it is pertinent to the function of the blood groups in man which are determined by multiple alleles.

Alleles are the alternative characteristics that a given gene may have. Since the chromosomes are paired, each individual would have two alleles for every characteristic that is controlled by a single gene. However, this does not mean that only two possible alleles exist for a characteristic. If, for example, sweet-pea color is a single gene characteristic, one plant could have only two alleles for color, let us say red (R) and white (r). But we know that sweet peas come in other colors; therefore there must be other gene alleles that produce these other colors if they are present.

This is the case in blood types. The gene is called (L) for Karl Landsteiner who discovered this system of blood groups. The (L) gene may have alleles A, B, or O. Each individual can carry only two of the three alleles and these alleles determine his blood type:

Name	Alleles
Type A	$L^A L^A$ or $L^A L^O$
Type B	$L^B L^B$ or $L^B L^O$
Type AB	$L^A L^B$
Type O	$L^O L^O$

The L^O allele acts like a recessive gene; L^A and L^B both dominate L^O; however, they do not dominate each other. (The same blood types are found in anthropoid apes, but not in monkeys.)

L^A produces antigen A, and L^B produces antigen B, but L^O does not produce an antigen. The antigens, when present, are carried on the surfaces of the blood cells.

One might expect that if blood type B were introduced into an individual with blood type A, antibodies would gradually be produced in the blood plasma in the usual antigen-antibody reaction, but this is not the case. The individual comes equipped with these particular antibodies at birth, possibly formed by direction of a separate gene. So instead of a gradual buildup against the foreign cells, a massive collision occurs. The individual's own Type A blood cells carry *A* antigen and the body has produced *b* antibodies, carrying them in the plasma for use against any "invading" Type *B* cells. The *b* antibodies quickly surround the *B* cells and agglutination occurs. The blood vessels become clogged so that circulation is impossible and death ensues.

We can diagram the possibilities of various blood transfusions and indicate whether or not agglutination will occur. This is shown in Fig. 76. We must keep in mind that since Type *O* does not produce antigens it will be the "universal donor" because it does not react with *a* or *b* antibodies present in the other types. Type *AB* has both antigens *A* and *B*, so it cannot be given to anyone with *a* or *b* antibodies. Type *AB* cannot have any *a* or *b* antibodies because it would coagulate its own cells; therefore, it becomes the "universal receiver." Since Type *O* does not have *O* antigens, there are no *o* antibodies.

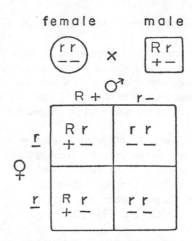

Fig. 77. Rh+ Male (Rr)×Rh− Female (rr)

	Type	Antigen	Antibody	Blood Type Donor				type
				A	B	AB	O	antigen
				A	B	AB	−	
Recipient	A	A	b	−	+	+	−	
	B	B	a	+	−	+	−	
	AB	AB	− −	−	−	−	−	
	O	O	ab	+	+	+	−	

Fig. 76. Occurrence (+) or Non-occurrence (−) of Agglutination in Blood Transfusions

One caution is always observed in transfusions. Although the blood types involved are the same, other factors in the blood may be incompatible. For this reason, the hospital always cross-matches the blood by mixing a droplet of the donor's blood with that of the recipient's. If agglutination or precipitation occurs, the transfusion is not made. The blood cells can be removed from the donor's blood, carrying away the antigen; then the plasma can be given without risk.

Among other known blood factors are the *M-N* types and the Rh-positive and Rh-negative types. The Rh factor was first found in *Macacus rhesus* monkeys and is now very well known. It is inherited as Rh-positive (Rh+) dominant over Rh-negative (Rh−) in a simple Mendelian pattern. The problem is that if an Rh+ male is mated with an Rh− female, the offspring could either be Rh+ or Rh−. See Fig. 77.

If the fetus (unborn baby) is Rh+, it will form Rh+ antigens. Under normal conditions, no blood passes between mother and baby; each has its own circulation, and fluids are exchanged through the placental tissue by osmosis. If a few of the baby's Rh+ cells escape into the mother's system in some way, the mother will form antibodies against the Rh+ cells. Antibodies normally pass through the placenta from the mother to provide the baby with protection from infection and disease until it is able to form its own antibodies. The harmful Rh+ antibodies will also be passed, but they do not usually increase enough during the first pregnancy to affect the baby. If a second pregnancy occurs, the antibody level is sufficient to affect the baby and cause agglutination. The baby may die before birth, but if it survives to term, a complete blood transfusion is necessary immediately at birth to remove all the antibodies and coagulated cells. A time interval of several years between pregnancies resulting from the union of an Rh+ male with an Rh− female is usually

Parent	Types	Offspring Possible	Offspring Not Possible
O	O x O	O	A, B, AB
A	O x A	O, A	B, AB
	A x A		
B	O x B	O, B	A, AB
	B x B		
AB	A x B	O, A, B, AB	
	O x AB	A, B	O, AB
	A x AB	A, B, AB	O
	B x AB		
	AB x AB		

Fig. 78

recommended to lessen the risk of antigen-antibody reaction.

Blood factors have been used to help determine parentage in legal cases or when it is suspected that babies have been mixed up in a hospital. The blood types cannot prove the identity of a suspected parent, but they can prove that certain people could not be the parent.

Fig. 78 shows the offspring blood types which are possible for various parental combinations.

For example, a court case showed that a mother was Type *AB*, Rh⁻; her infant son was Type *B*, Rh⁻. The alleged father was Type *A*, Rh⁺. Could he be the father? See Fig. 79.

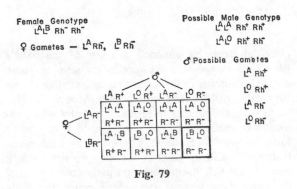

Fig. 79

We cannot prove whether the father is homozygous or heterozygous L^A or Rh⁺. If the father is heterozygous ($L^A L^O$ Rh⁺ Rh⁻) the child could be his, as shown in the lower right square.

Other special relationships occur between genes. A **complementary gene** is one which may be inherited separately but which requires the presence of its complement to produce the dominant effect. This separate inheritance factor was discovered when two kinds of white sweet peas were crossed and all the F_1 flowers were purple. Yet each white pea bred white when mated with its own kind. By crossing the F_2 hybrids it was determined that the purple color resulted only when two different dominant genes were present. Presumably one gene produced the raw materials in white while the other gene produced the enzyme which transformed the white material to purple pigment. Alone, neither gene could complete the sequence.

Supplementary genes are also independent. In this case, one gene will be dominant whether the other is present or not. The second gene, however, will not perform in the absence of the first.

A single gene may, at times, control more than one characteristic. In other instances a whole series of genes may be called into play in order to produce a single characteristic. Investigations into the latter type of action may be extremely difficult and complex.

Research in genetics such as the foregoing is still active but it is no longer considered the "new" genetics. It has, in fact, been referred to as "classical genetics" and the "Drosophila school" in recent years.

The tremendous frontiers opened by research in cellular biology and biochemistry have created an entirely new "new genetics." DNA structure, activating mechanisms for enzymes, enzyme inhibitors that "turn off" the enzymes at the proper stage—all these and many more features are being studied by geneticists.

POPULATION GENETICS

Mandelian genetics deals in terms of individuals, even though we are able to develop ratios that apply to large numbers of individuals. Groups of individuals form populations, and populations have a continuity that exists regardless of the fact that individuals die and are replaced. The name given to a unit of population is the **deme.**

Most of us are familiar with the word "species," the term used for the smallest unit group in identifying animals or plants that are closely related to each other. Linnaeus developed the system of binomial nomenclature in which each species receives two names, that of the genus to which it belongs, and that of the species itself. Because both names must be given, the system is called "binomial."

In general, a species is limited to those similar organisms that are able to interbreed freely. Members of a given species may live in different populations or demes. The heredity of the different demes is similar, but each unit may show different ranges of variation. These variations may intergrade and connect the demes.

If a population, by dividing asexually, has uniparental reproduction there will be little variation and the genetic makeup of parents and offspring will be the same. Uniparental populations are not common because of the genetic limitations. When many varieties are produced, the population has a better chance for surviving changes in environmental conditions although few of the variants may be able to adjust to the change. By means of sexual reproduction a biparental population is able to produce a much wider range of genetic variations than a uniparental population. Because of the events at meiosis, there are many more opportunities for mutant alleles to occur, with crossing-over and recombinations adding even more variety.

At first consideration it seems impossible to measure quantitatively the characteristics of a deme. One might also conclude that the smaller groups of variants would disappear, drowned out by the larger groups. However, neither of these impressions is valid, as we shall see.

A **gene pool** is the term given to the total alleles that occur anywhere in a given population. We may think of it as a kind of genetic alphabet soup. The number of times a given allele occurs in the gene pool is known as the **gene frequency.** We can determine the gene frequency by taking a population sample and counting it, or we can put together an experimental population of known gene frequency which we can allow to breed at random. By counting the resulting genes it is possible to find out what happened to the original proportions of alleles.

Let us assemble an experimental population of two thousand *Drosophila*—one thousand males and one thousand females. We will assume that they mate totally and at random. In our population, 36 per cent of the males have the genotype *AA*, 48 per cent carry *Aa,* and 16 per cent are recessive type *aa.* The female population has the same proportion. We can say then that the gene frequency is .36 *AA*, .48 *Aa,* and .16 *aa*. For simplicity, we will assume that egg and sperm production are equal. We can chart the gametes for a thousand males as follows in Fig. 80.

Population	Possible Gametes A	Gametes a	Total
36% AA	360	0	360
48% Aa	240	240	480
16% aa	0	160	160
100%	600	400	1000
gene frequency	.6	.4	1.0

Fig. 80. Calculation of Gene Frequency

If we treat the total female population as 1 and the total male population likewise, the results of crossing *Aa* with *Aa* are shown in Fig. 81.

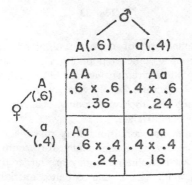

Fig. 81. Aa×Aa

We now have exactly the same gene frequency that we had previously: .36 *AA;* .24 *Aa* plus .24 *Aa,* or .48 *Aa;* and .16 *aa.*

This principle concerning gene frequency was first stated in 1908 by two men working independently— Hardy, a mathematician, and Weinberg, a physician —and is called the Hardy-Weinberg Law. It states that in a large population, with complete and random mating, and with no mutations, the gene frequencies will remain the same. In the example given above, gene *A* remains .6 and gene *a* remains .4.

Mathematically, we can state that the frequency F is equal to the quantity of the number of eggs containing A, $(p)A$ plus the number of eggs containing a, $(q)a$ mated to the quantity of the number of sperms containing A, pA, plus the number of sperms containing a, qa: $F = (pA + qa) \times (pA + qa)$. Multiplying by algebra we have:

$$\begin{array}{c} (p+q) \\ \underline{(p+q)} \\ p^2 + pq \\ \underline{+ pq + q^2} \\ p^2 + 2pq + q^2 \end{array}$$

Substitute:

$$pA = .6$$
$$qa = .4$$
$$\begin{array}{c} (.6A + .4a) \\ \underline{(.6A + .4a)} \\ .6A \times .6A + .6A \times .4a \\ \underline{+ .4a \times .6A + .4a \times .4a} \\ .36AA + .48Aa + .16aa \end{array}$$

If no mutations occur, if the genes are stable, and if mating is complete and random, the frequencies in the population will remain the same—at equilibrium. Since mutations do normally occur there will be a certain amount of change, but, in a large population, mutations will occur in both directions, $A \rightleftarrows a$, so equilibrium is maintained. If part of a population is isolated so that only a small "sample" reproduces, the direction of mutation may move, by indeterminate evolution, into a new and different state of equilibrium. This is known as **genetic drift.** We know that, in tossing pennies, we will get 50 per cent heads and 50 per cent tails in a large sampling. If we toss a sampling of only ten pennies, however, we may obtain only three heads and seven tails, or some similarly disproportionate ratio.

Total and completely random reproduction does not actually take place under most circumstances. Natural selection usually affects reproduction through various means. Those individuals who carry a defect will be prevented from breeding because they lack

Original Frequency			'A' Selected Against, Unable to Breed		
P₁ AA 36%	Aa 48%	aa 16%	P₁ AA 36%	Aa 48%	aa 16%

Gametes: A.6 a.4   Gametes: A.2 a.8

F₁

	A	a
A .6	AA .36	Aa .24
a .4	Aa .24	aa .16

F₁

	A	a
A .2	AA .04	Aa .16
a .8	Aa .16	aa .64

F₂ AA 36%	Aa 48%	aa 16%	F₂ AA 4%	Aa 32%	aa 64%

Fig. 82

the physical capacity to compete for a mate, or are unacceptable to potential mates. Thus, we have **nonrandom** or **selective mating.**

Let us once again use our population of 36 per cent *AA,* 48 per cent *Aa,* and 16 per cent *aa.* Now let us suppose that allele *A* becomes defective so that it occurs as a gamete in a frequency of only .2. The rest of the participating gametes will fill in so that the allele is raised to a gamete frequency of .8. See Fig. 82.

Naturally, one would not expect such a drastic shift to occur in one generation, for it is unlikely that a defect would suddenly affect 60 per cent of a population. The mechanism is present, nevertheless, for a rapid change to occur through natural selection. This is an important factor in the ability of a population to survive.

Variation is fundamental to life. Its source exists in bisexuality, in the behavior of the alleles at meiosis, and in the crossing-over and recombination of parts of chromosomes. Every individual varies from other individuals in some way, even if it is only slight.

When a single gene character determines appearance, as with red or white pea blossoms, we saw that a ratio of 3 : 1 resulted, an either-or choice, in the F_2 offspring. In the case of Andalusian chickens, with incomplete dominance, the ratio of F_2 offspring was 1 : 2 : 1. Where animals commonly occur in one of two or three definite forms, these types of ratios are always found. For example, some snail shells coil in a right-handed direction while a few coil in the opposite direction. This is called **polymorphism;** that is, the organism occurs in more than one form.

When more than one gene operates to determine

phenotype, as in the examples of height and skin color, the range of variation will be wider. There we found ratios of 1 : 4 : 6 : 4 : 1 that developed which could be plotted as a bell-shaped curve (Fig. 83).

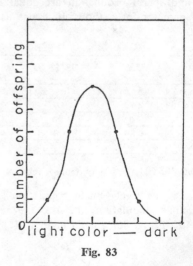

Fig. 83

The same type of curve is obtained if all the variations of a deme of population are plotted, provided conditions are constant for both environmental and genetic factors. The largest group on the curve, representing the most commonly occurring type, is called the **modal class.**

Environmental conditions may make a difference in the phenotype of the deme; for example, it is well known that children today are larger than their ancestors were. This is due to better nutrition, however, and apparently not to genetic change. If we plot a curve of the size of the ancestral population (P_1) and find the mean, we can then plot a second curve for the offspring F_1, which will indicate a shift of the mean toward the larger figure. But, since there has been no change in genotype, if the environmental conditions (nutrition) were to revert to those under which the parents grew, the succeeding generation F_2 would return to the same curve and mean shown by the P_1 generation. This is shown in Fig. 84.

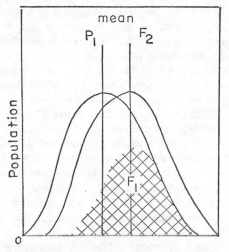

Fig. 85. Genetic Shift (F_1 and F_2 means are the same)

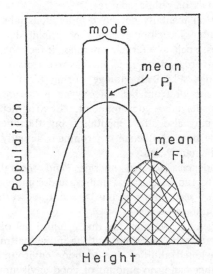

Fig. 84. Environmental Change

In the same way, we can plot a change in genetic conditions where environmental conditions are held constant. If we plot an original generation (P_1) we find the mean. If we then plot a succeeding generation F_1 and find a definite shift toward one extreme or other, we may have a genetic change taking place. If the mean for the next generation (F_2) remains at the F_1 mean, we will know that the change is a genetic one. See Fig. 85.

In living systems it is unlikely that either the environmental or the genetic conditions will remain absolutely constant. Therefore the amount of permanent shift in a generation will lie somewhere between the mean for the P_1 generation and the mean for the F_1 generation.

Variation that takes place within one deme will not necessarily be identical to that which takes place in an adjacent deme. If the demes interbreed to some extent there may, however, be no discernible difference. When the demes are scattered so that no interbreeding takes place each deme will establish its own characteristic gene pool and mutations. The differences between the demes tend to reflect adaptive variations, but nonadaptive variations definitely occur.

Several principles have been determined which predict the effect of environment on a single species. **Bermann's Rule** states that within one species the average size of the individuals tends to be smaller in warmer climates and larger in colder climates.

Gloger's Rule says that within any one species skin colors tend to be darker in warm, moist climates and lighter in cold, dry climates.

Allen's Rule specifies that within a given species tails, ears, bills, or other protruding parts tend to be shorter in colder climates than in warmer climates.

Differences between demes are shown to be correlated with environmental factors. In many cases the changes in temperature, elevation, moisture, and the like are gradual, not sudden. These are known as **gradients** or **clines.** As altitude increases on mountain ranges, a successive reduction in the size of certain plants can be measured. Geographical gradations are common, and are usually due to adaptive variations between the various demes.

EUGENICS

Living systems exhibit great capability for eliminating the defectives or extremes from the breeding population. If this were not so, the ability of the species to continue in existence would be severely jeopardized. Man has been the only species to attempt to nullify the natural mechanisms that eliminate the unfit. In so doing, man has, perhaps unwittingly, increased greatly the defective individuals who reach the age of breeding. In man this immediately becomes a philosophical and moral problem, as well as one of biological survival.

It is not our purpose to enter into a debate on ethical problems. It may be pointed out, however, that medical science has achieved unparalleled success in increasing both birth rate and survival in man. Many individuals are now able to reproduce whose defects a few years ago would have been lethal under former medical conditions.

In 1798 Thomas Malthus, the English economist and minister, saw that population increases are geometric, or multiples of themselves (2, 4, 8, 16, etc.), but that living space and food do not increase in the same proportion. He predicted that if all checks on population were overcome, the world would soon have a fantastic overpopulation coupled with severe want. In some countries this situation already exists.

For years, farmers and animal husbandry experts have steadily improved the quality of plants and animals by firmly establishing selective breeding as a principle. No farmer would deliberately seed his field with the poorest of his grain from the previous harvest, nor mate a prize cow to a scrubby, sickly bull.

Such problems, when encountered with regard to humans, are extremely controversial because of personal involvement. Historically, political and racial groups have tried to eliminate other opposing populations, sometimes under the guise of eugenics, "improving the breed." Making decisions about eugenic problems is not exclusively the responsibility of scientists and physicians. However, they must provide research and information which can contribute to the solutions that must be found for these pressing problems.

Many defects are inherited directly; in other cases, the tendency to develop certain deficiencies is inherited. Mental deficiency, or absence of mental development, can be an inherited characteristic. By definition, the types of defectives are: **idiots,** persons with a mental age of two years who are unable to give any self-care; **imbeciles,** a mental age of six; **morons,** a mental age of twelve. Morons may attain physical and sexual maturity and enter the general population. Since they are unlikely to succeed competitively in earning a living, they often become criminals or public charges.

Not all mentally defective individuals inherit their condition. Development may be impaired if certain viruses, such as German measles, infect the mother during pregnancy, or if she takes certain chemicals or drugs. Physical damage during birth, from instruments or from lack of oxygen for example, can cause defective development. Environment after birth may also affect mentality, but the question as to how much influence environment actually has remains unanswered.

Tendencies toward allergies and mental illness seem to run in certain families, and some groups or families seem more prone to contract certain diseases than others.

In general, we may say that a high level of mental ability may be inherited, but the individual may have undeveloped abilities due to poor environment. On the other hand, no amount of good environment will raise a congenital idiot above that level. The problems are to identify those children with underdeveloped abilities who could be helped by a change in environment and also to determine when a retarded individual has been developed to the limits of his capacities.

CHAPTER 7

ORGANIC MAINTENANCE

Living organisms are dependent upon their environment, which must continuously provide the substances necessary for survival. Earlier, we indicated possible ways in which the earth might have been formed and certain elements provided. These elements probably combined in many ways, eventually forming the necessary complex molecules that led to living systems.

Once living systems began to function extensively, the environment was in turn affected by them. This interplay has continued, and is an integral part of the life function. The living organism is in a constant state of change, and so is the environment. Balance is as important in the environment as stasis is in the organism because the physical world is constantly altered by the living world. Materials taken from the earth must be returned in some form, or the supply would eventually be depleted.

ELEMENTS REQUIRED BY LIVING SYSTEMS

Several cycles exist which illustrate the interrelations between the animate and the inanimate worlds. One of these is the **oxygen cycle** in which atmospheric oxygen is taken in by both plants and animals for respiration. In the organism, it is combined with hydrogen to form water. The water formed by animals may be excreted, or it may be broken apart again and the oxygen used in building proteins. In plants, by photosynthesis, the water is broken apart and the hydrogen is used to build carbohydrates; the oxygen is given off into the atmosphere.

Hydrogen is not used by plants or animals in its molecular form, the gas H_2, which is toxic to most living systems. Hydrogen released by metabolic processes in the cell is passed by a series of enzymes in what is called the **cytochrome system** until it can be combined with oxygen to form water. Oxygen is necessary in most living systems and is brought to the cell by diffusion and transport, as we shall shortly see.

The hydrogen ion (H^+) concentration, which determines the acidity or alkalinity of a solution, is expressed in chemical activities in terms of pH. Some compounds, when dissolved in water, dissociate or separate into free ions with positive or negative

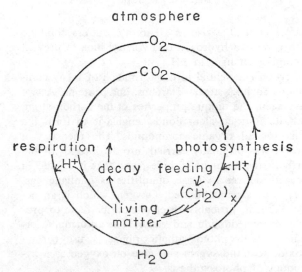

Fig. 86. Oxygen, Carbon, and Hydrogen Cycle Interaction

charges. Since the ions had been in compounds, the total positive and negative charges are equal. If, upon dissociation, an excess of hydrogen ions (H^-) form, the solution is **acid;** if more hydroxyl ions (OH^-) are formed, the solution is **basic,** or alkaline.

The pH scale is based on the number of free hydrogen ions compared to the number of free hydroxyl ions in a solution. Mathematically, pH$= -\log_{10} [H^+]$; the negative logarithm of the moles of H^+ ions present in a liter of a solution. This factor can also be measured by appropriate electrical meters. The scale ranges from 0 to 14, 7 being neutral.

We may think of this as a balance between the amounts of (H^+) and (OH^-). The numbers progressing from 7 toward 0 indicate a decrease in alkalinity and an increase in acidity, while numbers progressing from 7 to 14 indicate decreasing acidity and increasing alkalinity.

Many processes in living organisms produce excess acid or excess alkali. Therefore, an important system is that of the **buffers**—ions such as bicarbonate (HCO_3^-), carbonate (CO_3^{-2}), and phosphate (PO_4^{-3}), which can combine with excess (H^+) ions and prevent a damaging rise in acidity. Similar

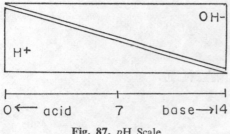

Fig. 87. pH Scale

buffers control a rise in alkalinity. Buffers combine with the extra hydrogen or hydroxyl ions to prevent damaging changes in pH.

Carbon is essential to the formation of many compounds such as sugars, starches, and proteins. As we have seen, the unique properties of the carbon atom, with its four covalent bonds, enable it to form the basis for all organic compounds. The theory that carbon was originally carried into the interior of the earth in the form of carbides, and that it existed in the atmosphere in large quantities as methane gas, CH_4, was also mentioned. Oxygen, which is an active element, probably reacted with methane to produce carbon dioxide and water. The autotrophic organisms with photosynthetic pigments use carbon dioxide to build sugars and give off oxygen as a by-product of photosynthesis.

Respiration was accomplished without free oxygen in the primitive environment, but carbon dioxide was probably the by-product, so more CO_2 was released into the air for photosynthesis. Carbon dioxide is also freed into the present atmosphere by the breakdown or decay of organic material. Atmospheric carbon dioxide is virtually the only source of carbon for living systems.

Nitrogen Fixation. Nitrogen is the most abundant gas (N_2) in the atmosphere close to the earth, but the majority of living organisms are unable to use it in this form. Nitrogen is essential for the formation of proteins and nucleic acids; yet organisms cannot take it from the atmosphere where it is plentiful. Despite the superiority of plants, which can trap sunlight for energy in photosynthesis, and animals, which can roam the world, both plants and animals are totally dependent on a few types of bacteria for their entire supply of usable nitrogen.

Nitrogen fixation is the process by which free atmospheric nitrogen combines with other elements to form organic compounds. Bacteria of the family Azotobacteriaceae have nitrogen-fixing ability when carbohydrates, such as cellulose, are present for energy. These bacteria secrete soluble nitrogenous compounds while alive, and, when dead, they are acted upon by decay micro-organisms—a process that releases ammonia.

Another group of bacteria, genus *Rhizobium,* are found as invaders on the roots of legumes, such as peas, beans, clover, and alfalfa. The plants form nodules on the roots in a typical reaction used to wall off foreign invaders but, because the adult plants are not able to use nitrogen or inorganic nitrates, they are dependent upon the secretions of the bacteria for usable nitrogen. The interrelation that develops is called **symbiosis** (living together)—each organism is dependent upon the presence of the other.

Nitrogen reduction, or denitrification, is accomplished by different bacteria. When plants and animals die, their proteins return to the soil where they are acted upon by many kinds of bacteria through hydrolysis, fermentation, and putrefaction (decay and rot). The proteins are broken down into amino acids; then, to ammonia. The wastes of animals consist of either ammonia (as in many of the invertebrates) or of substances such as urea and uric acid which can be hydrolyzed to ammonia. Some of the ammonia is released into the atmosphere or is denitrified further and released as nitrogen. Fortunately, the soil reacts to form ammonium salts which retain the nitrogen so that other bacteria can convert it into forms usable for plants.

Two groups of nitrifying bacteria perform nitrogen oxidation in two separate steps. One group, chiefly of the genus *Nitrosomonas,* oxidizes ammonium salts to nitrites ($—NH_4 \rightarrow —NO_2$). Nitrites are not usable by most of the higher plants and are even toxic to many organisms. Another group of bacteria, mostly of the genus *Nitrobacter,* oxidizes nitrite to nitrate ($—NO_2 \rightarrow —NO_3$). In the form of nitrate, soluble nitrogen may be taken into the roots of plants.

Animals have no means of obtaining the nitrogen necessary to build proteins for themselves. They are completely dependent upon plants which are, in turn,

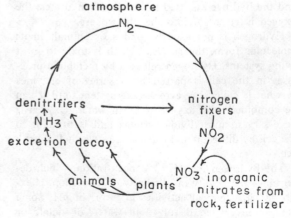

Fig. 88. Nitrogen Cycle

dependent upon bacteria. It is true that some animals do not feed on plants, but only on other animals; somewhere in the succession of feeders, however, is the animal which did feed on plants to obtain nitrogen.

In addition to the "big four" elements—oxygen, hydrogen, carbon, and nitrogen—that are used to build sugars and proteins, other elements are required by living systems. These elements are not necessarily the most abundant ones found in the crust of the earth, as can be seen readily by comparing the following lists.

Earth Crust Element % by weight		Living Matter Element % by weight	
1. Oxygen	47	1. Oxygen	62
2. Silicon	28	2. Carbon	20
3. Aluminum	8	3. Hydrogen	10
4. Iron	5	4. Nitrogen	3
5. Magnesium	2	5. Calcium	2.5
6. Calcium	3.6	6. Phosphorus	1.1
Others	6.4	Others	1.4

Although the elements listed above constitute the bulk of living materials, there are a number of others that do not occur in quantity but that are also required for life. These include sodium, chlorine, potassium, iron, copper, iodine, sulfur, zinc, magnesium, manganese, molybdenum, cobalt, and, in some cases, vanadium and selenium. Sodium, potassium, and chloride ions are important in osmosis, in buffer systems, and in nerve impulse transmission; sulfur participates in protein linkage. The so-called "trace elements" are generally employed as cofactors or structural parts of enzymes. (The general term cofactor is given to various types of substances that are required to activate an enzyme which will catalyze a given reaction. Without certain trace elements as cofactors, some enzymes remain inactive.)

DIFFUSION

Two physical principles or processes are important in understanding cell and body function—one is diffusion and the other is osmosis (which is really a special type of diffusion).

Molecules of all kinds are constantly in motion, even within solids such as iron and wood. The amount of motion reflects the kinetic energy of the molecules at various temperatures. We have seen that substances are solid at low temperatures, liquid at higher temperatures, and gaseous at still higher temperatures. An increase in temperature, hence in kinetic energy, causes the molecules to bounce around and separate themselves from each other.

For this reason, molecules move from an area of high concentration into an area of lower concentration. Diffusion is at work when, for example, perfume sprayed into the air gradually dissipates. If the molecules are restricted to a container, they will diffuse until they are evenly distributed; they will not stop moving then, but further random movements will not alter their concentration in any given area.

This occurs when we place a lump of sugar in a cup of coffee. The sugar molecules will pass (very slowly if the coffee is not stirred) from the area of concentration into an ultimate uniform distribution. If we add cream, we can see that it diffuses independently through the solution, regardless of the sugar. See Fig. 89.

Fig. 89. Diffusion

Diffusion is an important means by which oxygen and nutrients are introduced into cells. Because diffusion is a slow process in terms of distributing molecules throughout a relatively great volume, the size of an organism that depends on simple diffusion through its surfaces for its oxygen or nutrient supply is severely limited. Larger plants and animals must develop systems to bring external media in contact with cell surfaces. Gills on fish and crustaceans are constructed so that they can be "fanned." More water can be circulated across the gills, which are often "feathery" to increase surface area, increasing the amount of oxygen brought in contact. In man, the lungs are subdivided into many tiny sacs (alveoli) to increase the surface exposed for diffusion. A muscular chest cavity and diaphragm also expand and constrict the lungs, thus providing a circulation of air that contains more oxygen.

We know that in respiration not only is oxygen taken into the organism, but carbon dioxide is given off. Carbon dioxide, CO_2, is a waste product of cellular respiration in which carbohydrates are broken down to release energy. The fact that CO_2 is expelled while oxygen, O_2, is brought in points again to the principle of diffusion.

In human respiration, diffusing molecules pass from the higher concentration to the lower concentration until they become uniformly distributed. This is called a concentration or tension gradient.

In the air at the lung surface there is a high tension of oxygen compared with that inside the lung cell, so oxygen passes into the cell. The tension of CO_2 inside the cell is high compared with that of the air in the lung, so the CO_2 passes out of the cell into the air.

Transportation is necessary for the delivery and pickup of oxygen and carbon dioxide, as well as of other nutrients. The blood fluid is formed of water, protein, and inorganic ions. Blood cells contain hemoglobin, a pigment which can form a weak bond with oxygen. We saw earlier that the structural form of heme, one constituent, is very much like that of chlorophyll, except that heme has an iron center whereas chlorophyll has a magnesium center. Globin, the other constituent, is a protein.

The epithelial cells of the alveoli in the lungs and the thin walls of the tiny blood vessels (capillaries) lie in close contact. The gases diffuse between them easily, depending on the tension gradient. In man, the capillaries merge from the lungs to form the pulmonary veins that lead to the heart. The heart pumps blood out through the muscular aorta into the arteries which distribute it throughout the body. The hemoglobin (Hb) is red when combined with oxygen (HbO_2) in arterial blood, but when it gives up oxygen it becomes dark, as seen in venous blood. Veins return the low-oxygen venous blood to the heart where it is pumped through the pulmonary arteries to be aerated again in the lung capillaries.

Carbon dioxide can combine with hemoglobin, but most of it combines with water in the blood and is carried as bicarbonate ions:

Hemoglobin also combines with carbon monoxide —more firmly than it does with oxygen. Unvented gas heaters use up the oxygen in a closed room so that insufficient oxygen can combine with the carbon in the gas. A person subjected to carbon monoxide from an unvented gas heater will lose consciousness and possibly die because the hemoglobin in his blood is not carrying oxygen to the tissues. Ordinarily, combustion occurs with adequate air supply: $C + O_2 \rightarrow CO_2$. Lack of oxygen causes combustion thus: $2C + O_2 \rightarrow 2CO$. In automobile engines, the chambers are not supplied with enough oxygen to oxidize the carbon to CO_2; therefore, a leaky muffler can be dangerous to the occupants of a closed car if the engine is left running.

The nervous system in man meters the amount of CO_2 (as $HCO_3{}^-$) in the blood. The medulla oblongata region of the brain, at the base of the skull, influences respiration. It sends impulses to the diaphragm through the phrenic nerves which stimulate an increase in breathing if the CO_2 level rises.

There is a common misconception that the lungs squeeze air out through contraction and expand to bring air in—as though the lungs were muscular. The chest cavity is closed; it is the chest muscles that cause the chest to expand while the diaphragm contracts, producing a negative pressure in the chest cavity. Air then flows into the lungs and the alveoli expand. When the chest and the diaphragm relax, the chest cavity becomes smaller and air is pushed out of the lungs and the alveoli. The diaphragm is curved upward when in a relaxed position and is pulled down into a horizontal position on contraction.

OSMOSIS

Osmosis, an extremely important physical process, is actually a specialized example of the principle of diffusion. In osmosis, a membrane is present which may act as a barrier to the energetic motions of diffusing molecules. Each cell is surrounded by a membrane which is called the **plasma** or **unit membrane.** It is a double-layered lipoprotein having a gridlike structure with minute openings in it. It would seem that these openings would limit the size of the molecules which could pass through. However, proteins have charges on their terminal ends which may repel molecules having like charges. The lipids also are oriented so that one group attaches readily to one kind of molecule and another group to a different kind of molecule. These charges and

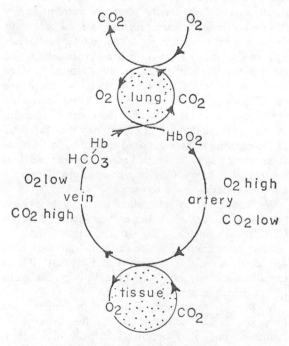

Fig. 90. Oxygen Exchange

reactions combine to make membranes differentially permeable or semipermeable.

If we select an artificial membrane it will have minute pores of a certain size in it. Suppose that we take a sack made of cellophane membrane and fill it with a colored sugar solution and suspend it in a beaker of water. If the pores are large enough, the beaker of water will soon contain colored sugar water with a concentration equal to that of the fluid remaining in the sack. The dissolved sugar solute has diffused through the membrane. The solution consists of a **solvent,** water, in which the sugar molecules are dissolved; this is known as a **solute.** When the entire solute diffuses through the membrane, the process is **dialysis.**

Dialysis can be used as a means of removing one substance from a solution, without removing another, by using a selectively permeable membrane. This technique is the basis for the artificial kidney machines, where it is necessary to remove waste products from the blood without removing blood cells. The blood is pumped from the patient through tubes of dialyzing membrane that are suspended in a water medium, then back into the patient. The tubes are usually filled with blood taken from the patient previously to "prime the pump" and to prevent a drop in the patient's blood pressure.

Osmosis occurs when the permeability of the membrane is restricted so that only the solvent molecules, and not the dissolved molecules, will pass through it. For example, let us suspend a membrane sack of colored sugar water in a beaker of water. This time we have chosen a membrane that has much smaller pores and the sugar molecules are unable to pass through, although the solvent molecules (water) are able to pass. The energies of all the molecules have not changed so they are still bouncing around. The difference is that the water molecules can pass through the membrane in both directions but the sugar cannot. In an attempt to equalize the concentration on each side of the membrane, the water will, in effect, go in the direction of the higher concentration of the sugar (or of any salt). The level of solution in the sack will rise and

will continue to rise up a tube inserted in the closed sack. This is shown in Fig. 91.

This phenomenon is **osmotic pressure** and is extremely important in biological systems. Since cells are composed largely of water, the substances in the cell, such as sugars and salts, determine the osmotic pressure in each cell. The blood and other body fluids must maintain a balance with the cells at all times. When the osmotic pressure within a cell is equal to the osmotic pressure of the fluid surrounding it, equilibrium occurs and the fluid is said to be **isotonic.** If the fluid contains more salts than the cell, it is **hypertonic,** and water will pass out of the cell, causing it to shrink. If the fluid contains fewer salts than the cell it is **hypotonic,** and water will pass into the cell, causing it to swell.

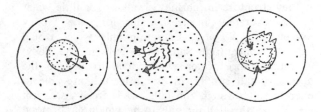

isotonic   hypertonic   hypotonic

Fig. 92. Red Blood Cell in Salt Solution

The net result of osmosis is that **water will move in the direction of the higher concentration of salts.** If we add salt to whole blood, the blood cells will shrink or wrinkle up. If we dilute the blood with water, the cells will expand until they burst. A common reminder of osmosis occurs when plants are fertilized. The instructions with commercial fertilizers are to dilute the fertilizer or to water heavily. If this is not done, the treated plants may suddenly turn brown and die, and we say that the fertilizer "burned" them. Normally, ground water flows into the plant where the concentration of salts is higher. When we add salts (fertilizer) to the root area, the water flows out of the plants into the ground salts; in effect, the plants then "water" the ground.

Obviously, a fish swimming in a fresh-water lake would swell up and burst as a result of osmosis if it did not have some means of pumping water out or salt in. Many cells, perhaps all, have the ability to "pump" actively against the osmotic gradient, and many are able to conserve or concentrate salts. Enormous amounts of cell energy must be expended to "pump" against the gradient.

While aquatic animals and plants may have to cope with too much water, terrestrial plants and animals must guard against water loss. The excretory devices of all animals allow for this factor.

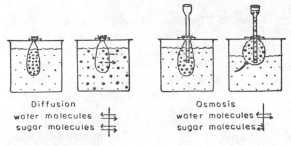

Diffusion
water molecules
sugar molecules

Osmosis
water molecules
sugar molecules

Fig. 91

In the lower animals, it is thought that the excretory device is purely for controlling excess water. Some protozoans have a contractile vacuole, an area in the cell into which water seems to drain, collect, and from which it is then squeezed out by a sudden contraction. Some of the lower invertebrates have ducts called **nephridia** which consist of a ciliated funnel structure with a tube leading to the outside. Waste products of metabolism exit through the nephridia. Eggs and sperm that might be released by the gonad inside the body could also exit through the excretory pores. Coelomic ducts from the body cavity to the exterior probably play a similar role in higher animals.

In sea water, where the concentration of salt is higher than it is in the fish living in the water, the fish must have mechanisms to retain water. The kidney structure in marine fish allows for little water to be excreted; in fresh-water fish, the kidney excretes more water. Excreting little water would tend to concentrate salts excessively in the body; excreting more water would tend to deplete the necessary salt supply. Most marine fish are able to excrete salt through the gills, while fresh-water fish absorb salt through the gills to help maintain osmotic equilibrium.

The method by which individual cells accomplish pumping against the osmotic gradient is not clear. There is some evidence that pinocytosis (cell-drinking), which was described earlier in the section on cell structure, plays a large part in active transport. The cell membrane and cytoplasm put out projections, or podia, and surround the fluid to be taken in. Authorities now feel that particles located between the two layers of the cell membrane rotate to provide pumping. Electron microscopy shows that this is possible.

In terrestrial (land-dwelling) animals, the body must conserve water. The kidney of vertebrates is, therefore, constructed so that waste materials may be filtered out of the blood while most of the water remains in the system. This means that the urine will be concentrated, or hypertonic. Other mechanisms to protect against water loss exist; for instance mucus is secreted by amphibian skin; scales are seen on reptiles; and, even in man, the outermost layer of the skin is **cornified**—composed of dead, hardened epithelial cells.

METABOLISM

One of the distinguishing characteristics of living organisms is metabolism—the process of converting food into energy or into the building of new tissues by means of complex chemical reactions.

There are a number of ways by which the various types of foods—sugars, fats, and proteins—are converted by the body into energy and tissue. We have mentioned the need for carbon, hydrogen, oxygen, and nitrogen and also the need for certain other elements that may be incorporated in food or used in small quantities as catalyzers. In addition to those elements that we have already mentioned as required by organisms, there are a number of more complex substances that also are essential.

Amino Acids. Organisms are generally capable of using many different substances as food, breaking them down to their components, and synthesizing those needed by the body. There are some substances, however, which certain animals are unable to make for themselves and which must be incorporated in the diet ready for use. These include some of the so-called essential amino acids. About twenty amino acids are considered essential and almost half of them cannot be made by the living body, apparently because it lacks the ability to build certain carbon ring structures from available raw materials. It is possible that at one time in biological history the organisms may have been able to build the necessary amino acids, and they may have lost this ability through mutations.

Vitamins. Amino acids are required by organisms to build proteins. There are other substances essential for life which are not required as food, but which act as coenzymes or cofactors in metabolism and are required in very small quantities compared to the essential amino acids. These substances are called vitamins and must be supplied to those organisms which cannot synthesize them. A substance may be classed as a vitamin for one group of animals while another group may be able to synthesize the same substance, avoiding the need to be supplied with it. Consequently, animals can obtain vitamins they lack by eating plants or animals that have synthesized those particular substances.

In man, the required vitamins are not all chemically similar. They may be divided into two groups: the **fat-soluble vitamins** and the **water-soluble vitamins.** A brief summary of the sources and clinical effects of the vitamins is given below, although we shall not go into the chemical structures to any great extent.

Fat-soluble Vitamins. Vitamin A is an alcohol, found only in animal products such as butter, eggs, and fish liver. Some plants possess **carotenes,** yellow pigments that act as **provitamins**—substances that can be converted into vitamins. Animals that eat these plants can convert the provitamin carotene into vitamin A.

Vitamin A is necessary to maintain normal secretions of epithelial tissues, such as those of the skin, eye, and digestive and respiratory tracts. It is known as the anti-infection vitamin. Vitamin A-deficient rats show coarse dry fur and scaly red eyes. Bone and tooth development is retarded; extreme nervousness is seen. Night blindness in humans is due to a deficiency of vitamin A.

Vitamin D is actually a group of sterols, D, D_2, and D_3; it is found primarily in fish livers and in the livers of animals that feed on fish. It is frequently added to milk, oleomargarine, and other foods to enrich them. The D_3 provitamin is deposited under the skin in man and ultraviolet rays from sunshine activate it.

The role of vitamin D in preventing rickets has long been recognized. This disease results in abnormal skeletal development due to lack of calcium. Recently, the role of calcium in general metabolism has been examined. Calcium is frequently associated with phosphorous in the body, and phosphates provide much of the energy required for cell function. Therefore, vitamin D may be more important than was previously thought.

Overdoses of both vitamin A and vitamin D can be toxic and should be avoided. Excess vitamin D may lead to calcification of tissues which are supposed to be soft, such as those in the lung and kidney. This excess is especially dangerous in pregnancy.

Vitamin E includes three substances known as **tocopherols.** No need for these has been established for man, but in rats a deficiency causes, in females, the resorption of the fetus in the uterus and in males, sterility. Certain types of muscular dystrophies are seen in E-deficiencies in some animals, but humans do not seem to respond to E therapy. Some researchers feel that tocopherol acts in electron transport in the cytochrome system, with selenium as a trace element cofactor, in cellular respiration. Vitamin E is found in so many foods common to man's diet that no special effort to include it is required under normal circumstances. It occurs in leafy vegetables, milk, eggs, meat, and wheat.

Vitamin K is best known for its role in preventing hemorrhage. It is apparently necessary in order for the liver to produce prothrombin, the precursor substance of thrombin, which converts fibrinogin to strands of fibrin in the clotting of the blood. In addition to this obvious activity, one of the K vitamins is necessary as a cofactor in plants and animals for oxidative phosphorylation, the energy-producing breakdown of sugars in which phosphate groups transport the energy.

Vitamin K is found in green leafy plants. The principal source, however, is from intestinal bacteria which secrete vitamin K. Newborn infants are lacking in intestinal "flora" at birth and may be subject to hemorrhage if the mother has been lacking in the vitamin. Sometimes antibiotics taken during virus attacks will destroy the intestinal bacteria and cause a temporary tendency to hemorrhage. Molds and yeasts, which are low in vitamin K production, may flourish when the bacteria normally present are destroyed, making it difficult to re-establish a balance of organisms.

The supply of the fat-soluble vitamins in man is dependent upon the ability of the intestine to absorb them. Bile, which is secreted by the liver, is stored in the gall bladder and carried into the small intestine by the bile duct when food is eaten. The bile emulsifies fats into small droplets so that the enzyme lipase can split the molecules. If the bile system is not functioning properly, fat-soluble vitamins will pass through the body unused. Mineral oil, which is sometimes taken to relieve constipation, is indigestible and may block the absorption of vitamins. Occasionally, a manufacturer develops a "nonfattening" salad dressing with a mineral oil base; this, too, will prevent the absorption of essential vitamins.

Water-soluble Vitamins. The water-soluble vitamins include vitamins C and the B complex. Vitamin C, ascorbic acid, is known to be a reducing agent so it is probably involved in hydrogen transport. Vitamin C is necessary to maintain supportive tissues such as bone, cartilage, and connective tissue. Scurvy, the vitamin C deficiency disease, causes loosening of teeth, increased bone breakage, and poor wound healing.

Man and other primates and guinea pigs are the only animals that are unable to synthesize their own ascorbic acid; this inability is probably due to an inherited gene mutation. Ascorbic acid is found in citrus fruits, melons, tomatoes, raw cabbage, and salad greens. It is destroyed by cooking and by storage in some metal containers or chopping with metal instruments.

Years ago, sailors on long voyages without fresh fruits commonly suffered from scurvy. When it was discovered that citrus fruits prevented scurvy, the British navy required the sailors to include limes acquired in Mediterranean ports in their diets. Hence, the British sailors became known as "limeys." The earliest cure for scurvy was found in Jacques Cartier's records of his expedition to Canada in 1536, where Indians prescribed an extract of fir needles for the men.

Originally, the term "vitamine" described a substance, found by the biochemist Casimir Funk, the lack of which was responsible for another nutritional disease—beriberi. This substance was eventually named "vitamin B" when other workers found water-soluble factors besides Funk's that produced similar symptoms when they were absent from the diet. Beriberi is characterized by muscular fatigue and weakness followed by painful degeneration of the nerves and paralysis.

Vitamin B_1 is thiamine, found in small quantities in yeast, liver, nuts, pork, and whole grain cereals. Since cereals are usually refined, thiamine is added to bread and cereals to enrich them. Thiamine functions in the body as thiamine pyrophosphate, a coenzyme in carbohydrate metabolism. Beriberi is a deficiency disease found in countries where rice is the main diet and where the thiamine is removed when the rice is polished.

Vitamin B_2, riboflavin, is a yellow-green pigment found in milk, liver, kidney, heart, and vegetables, and in anaerobic-fermenting bacteria. It may be destroyed in milk during processing, by light, and by irradiation for vitamin D. In cooking the meat of organs, 70–80 per cent of vitamin B_2 is retained. In humans, deficiency in riboflavin is characterized by cracking at the corners of the mouth, purplish tongue, and loss of hair. It is relatively common in mild forms; severe deficiency can result in death.

Flavins are important in forming flavoprotein enzymes, which are involved in electron transport in the cytochrome system. FAD, flavin adenine dinucleotide, is a coenzyme involved in cellular oxidations.

Niacin, or nicotinic acid, which is found in meat, yeast, fresh vegetables, and beer, is the anti-pellagra vitamin. Pellagra is a disease characterized by inflammation of the epithelial tissues of the skin and digestive tract. Inflammation of the nervous system may cause mental upsets.

The amino acid tryptophan is normally converted to niacin by the body so that diets low in protein are generally niacin-deficient. In regions where corn is the staple food, pellagra is common because corn has a low niacin content. The intestinal bacteria in animals synthesize a large quantity of niacin. These bacteria may be destroyed when antibiotics are taken, and their loss may result in niacin deficiency. Niacin is a constituent of coenzymes I and II, which are extremely important as hydrogen and electron acceptors in the cellular energy cycle. Coenzyme I is nicotinamide-adenine-dinucleotide, NAD (formerly called DPN). Coenzyme II is nicotinamide-adenine-dinucleotide-phosphate, NAD·P (formerly TPN).

Vitamin B_6, pyridoxine, is found in a variety of plants and animals and particularly in cereal grains. It occurs in three forms but apparently is used in the form of pyridoxal phosphate in a variety of roles, such as catalyzing amino acid metabolism and for the transfer of amino and sulfur groups. Because B_6 is found in many foods, true deficiencies were not at first identified. It now seems to be essential in the nutrition of all animals investigated so far.

Deficiencies of B_6 have produced severe anemias. Neural symptoms similar to epilepsy have been identified in infants who were fed an unsupplemented formula that was sterilized at a high temperature under pressure, which apparently destroyed the B_6. Degeneration of the central nervous system has been demonstrated due to a lack of this substance. Intestinal bacteria contribute also to B_6 synthesis.

Pantothenic acid is necessary for the maintenance of so many tissues in so many plants and animals that it is obviously an important factor in cell metabolism. It is a component of coenzyme A, which is active in metabolism of acetate, participating in the final breakdown of carbohydrates. It is also involved in forming the substance acetylcholine, which is essential to the transmission of nerve impulses, and in the formation of the oxygen-carrying substance heme. Acetate is a precursor to the formation of the hormones.

Pantothenic acid is found in quantity in egg yolk, kidney, liver, and yeast, and in lesser amounts in broccoli, lean beef, and molasses. Deficiencies are known to cause dermatitis and neuritis (skin and nerve irritations), gray hair, and also the breakdown of the adrenal gland, which forms the hormones adrenaline (epinephrine) and norepinephrine.

Biotin was first discovered as a necessary growth factor in yeast. (A large proportion of the studies of cell metabolism are performed with micro-organisms.) Biotin is normally produced by intestinal bacteria and is widely distributed in foods so that a true deficiency is unlikely. However, raw egg white contains the protein avidin which completely blocks absorption of biotin in the intestine. A fancier of a raw egg diet might, therefore, show biotin deficiency symptoms of dermatitis, falling hair, and muscle weakness. Cooking eggs destroys the egg-white factor.

Folic acid is composed of at least three slightly different chemical substances with different activities. They function as coenzymes in the metabolism of carbons to form some amino acids and probably DNA. Consequently, they are essential to rapidly reproducing cell systems. For this reason folic acid deficiency affects the production of red blood cells and anemia results. No doubt the deficiency affects other tissue systems in the same fashion, but the anemia symptoms of fatigue and low red blood cell count are easily observed.

Para-aminobenzoic acid is also a growth factor for micro-organisms. It appears to be necessary for the supply of molecular structures needed to build the folic acid molecule in those organisms capable of synthesizing folic acid.

Vitamin B_{12} has been given dramatic attention in the past twenty years for it has proved to be the cure for pernicious anemia, a debilitating, often fatal form of anemia.

The American zoologist William Ernest Castle postulated the existence of two substances important to the prevention of anemia—an intrinsic and an extrinsic factor. Castle was proved correct in the sense that external food sources such as milk, meat, and eggs normally provide B_{12}. It has been discovered that the intrinsic factor involves the presence of hydrochloric acid and a gastric mucoprotein found in the stomach. If the acid factor is not present it is impossible for the intestinal mucosa to absorb vitamin B_{12}. From one point of view vitamin B_{12} is both the intrinsic and extrinsic factor, because if it is injected the anemia responds. The administration of gastric juice or dilute hydrochloric acid along with normal sources of B_{12} will also cause remission of anemia if it is due to inhibition of gastric secretions.

Other vitamins or coenzymes will undoubtedly be discovered in the future. As more has been learned about metabolism at the cellular level, it has become apparent that a highly complex interplay is involved in maintaining the functional balance that is taken for granted in the plant and animal world.

DIGESTION

The system of food intake and of digesting the food ingested depends on the general level of complexity of the organism and its particular diet. The unicellular animal performs all of its functions within the confines of the single cell—**intracellular digestion.** This mode of digestion extends into the multicellular animals on the most simple levels where it is likely to be "every cell for itself" in digestion. Gradually, specialization became necessary for the survival of multicellular animals. It was more efficient for an organism to have an outer covering of cells adapted for protection and perhaps for trapping food, while more delicate secretory cells performed digestive functions for all the other cells. Thus, **extracellular digestion** was initiated. These cellular specializations must logically be accompanied by a transportation system for passing the nutrition around, and by a co-ordination system for regulating the functions.

Cells are organized into tissues and eventually tissues are assembled into organs. The progressive increase in complexity seen in animals is paralleled by the complexity of the organ systems that have developed. The digestive system progresses from group to group of animals, changing from a simple sac with two-way traffic into a more efficient one-way tube or gut.

Folds in the surface of the gut and the growth of diverticula helped to solve the problem of increasing the surface to provide nutrition for increased animal size. Specialization within the digestive tract progressed according to the needs of the animal; apparently it is more efficient to divide the tract into separate interconnected compartments. We can see these adaptations in the digestive tract in man, where separate structures secrete specific enzymes and a particular pH can be maintained which may differ from that in another digestive area that is doing a different job.

In man, the mouth is equipped with teeth, tongue, and salivary glands. In the mouth, food is reduced to manageable size and is mixed with an enzyme, salivary amylase (ptyalin), at a neutral pH. The principal chemical action converts starch to maltose.

The food, gathered in a **bolus,** a soft mass, passes next into the esophagus and is propelled down to the stomach by muscular contractions. A sphincter muscle (circular closing ring) opens into the stomach when the bolus reaches it, allowing it to pass through.

Stomach. The stomach, lying just beneath the lower left ribs, is thick-walled and muscular. It is lined with mucous membranes which secrete enzymes and gastric juice containing a large amount of 0.2 to 0.5 per cent hydrochloric acid. Here the food is churned by muscular contraction until it is thoroughly exposed to the enzymes pepsin and rennin, making it a creamy consistency. Rennin acts mainly in coagulating milk, while pepsin breaks proteins down into smaller segments called proteoses and peptones. In order to prevent pepsin from digesting the stomach walls themselves, the enzyme is secreted first in an inactive form and is activated by the acid released when food enters the stomach. Advertisements offering products to "neutralize excess acidity due to overeating" are possibly doing the stomach no favor, since the acid pH is essential to digestion. Instead, the carbonates introduced in such medication tend to interact and the gas bubbles released help reduce pressure, but do not necessarily speed digestion.

Small Intestine. The creamy mixture called **chyme** is released into the small intestine by the pyloric valve at the bottom of the stomach. The acid pH stimulates the release of several hormones by the duodenum, the section of small intestine leading from the bottom of the stomach. These hormones, collectively known as secretin, are absorbed by the blood vessels of the intestinal wall. Secretin and pancreozymin are carried to the pancreas, hepatocrinin to the liver, and cholecystokinin to the gall bladder, and they all serve to stimulate these digestive glands. Enterocrinin stimulates the flow of intestinal juices. The glands arise embryonically as diverticula of the gut, demonstrating the evolutionary trend to the increased surfaces and specialization of areas mentioned earlier.

Pancreas. The pancreas secretes a fluid that is highly alkaline and contains a number of inorganic ions such as sodium, potassium, chloride, and bicarbonate. The pancreatic enzymes include some that affect protein such as trypsin, chymotrypsin, and peptidases. Lipase acts on fats; ribonuclease and amylase act on nucleic acids and starches respectively. Some of these enzymes are inactive precursors, called zymogens, that must be acted upon by the secretions in the intestine before they become functional. Trypsin, for instance, is activated by enterokinase which is produced in the intestinal glands. The pancreatic duct opens into the duodenal area, as does the common bile duct from the liver and gall bladder.

While the pancreatic enzymes act on proteins, sugars, and fats, bile produced in the liver acts as a detergent and emulsifier. By lowering surface tension and making fats water-soluble, fats may be digested and vitamins carried by fats may be released for absorption. Bile also is quite alkaline and helps to maintain the proper pH in the intestine.

The secretions of the small intestine include aminopeptidase and carboxypeptidase; the latter acts on peptide bonds to produce free amino acids. Maltase, sucrase, and lactase each split specific bonds in the various sugars to produce glucose and fructose. The enzymes, in general, have specific actions, and are effective only on a single bond in a sugar or protein. The lack of a certain enzyme means that a particular foodstuff will pass through the body unused.

Most of the absorption of the digestive end-products occurs in the small intestine. Some of them are transported by osmosis through the cells of the intestinal villi into the blood vessels in the centers of the villi. Other substances that cannot diffuse against the gradient must be actively transported by the cells and require considerable energy for this purpose.

Fats pass from the villi into the **lymph ducts,** which are larger channels distributed through the body. These vessels unite to form a channel that empties into the circulatory system through a large duct located beneath the left collarbone.

Large Intestine. The large intestine serves primarily to extract water from the intestinal contents in order to conserve the water supply and to compact the solids (feces) for evacuation.

INTRACELLULAR DIGESTION

Formerly, a discussion of the digestive process ended when the delivery of the end-products to the cells by the blood was described. During the last thirty years, however, investigations of the activities within the cells themselves have led to some understanding of what happens to molecules after they are delivered to the cell. The story is exceedingly complex and involves some knowledge of organic chemistry and biochemistry. We shall attempt to examine the principal aspects briefly without becoming overinvolved in chemical reactions.

The liver is the primary site of chemical action for the contents of the blood coming from the intestinal villi via the hepatic portal vein. The liver, which probably evolved from a simple digestive gland, carries on a number of chemical processes. It converts glucose, fructose, and other hexose sugars ($C_6H_{12}O_6$) into glucose phosphate and then into glycogen by splitting out water to link the sugar units. Glycogen is the "storage product" of the liver. A certain level of glucose is constantly required in the circulating blood to nourish other cells and to furnish energy. The liver can convert glycogen back into glucose when the glucose level in the blood drops. Glycogen is also stored in muscle cells as a reserve for muscular activity.

The liver detoxifies certain substances by altering their chemical structures into harmless forms. This includes controlling ammonia, which is produced in the intestine by protein digestion and bacterial activity. The liver also retrieves some of the important ions when the red blood cells break down, and it manufactures substances vital to blood coagulation.

The cellular metabolism of glucose provides energy for the activities of the cell and occurs in steps both in the cytoplasm and in the mitochondria along the surface of the folds called cristae. The process occurring in the mitochondria is called cellular respiration because oxygen is used and carbon dioxide is given off at the end of the sequence. There are many separate reactions, each one of which may be reversible. There are also a number of choices at certain steps—options are available as to which metabolic pathway will be taken.

Bond Energy. In order to understand why various chemical changes produce energy, and are not simply "chemical juggling acts," we must consider the concept of bond energy. We have already seen that electrons circle around the nucleus of an atom in various orbits, and that, in the formation of compounds, the distance the electron "falls" from its orbit determines the amount of energy released. The energies of different kinds of bonds differ as well. You recall that the hydrogen bond was referred to as a weak bond while covalent bonds are much stronger.

The energy between various kinds of bonds has been measured; for example, the covalent oxygen-hydrogen (O-H) bond energy is 220.3 kilocalories per mole, while that of a hydrogen bond between oxygen and hydrogen (O··H) is only 4.5 kilocalories per mole. The units need not concern us; note only the comparison of the energies.

Energy must be put into creating the bonds, and it is the energy of sunlight that combines carbon dioxide and water into sugars by means of photosynthetic pigments in plants. Breaking these carbon dioxide-water bonds in the cell will release this energy.

Energy is never created or lost; it is only changed in form. During a reaction, for example, some energy may be given off as heat, some may be translated into motion, or some may be transferred into a new compound with a different amount of energy potential.

We know that sometimes a compound may exist in two different forms; these are called **isomers.** For example, there are right-handed sugars and left-handed sugars that have exactly the same numerical formula but structurally are mirror images of each other.

Some compounds do not have stable electronic structures; consequently, the electrons migrate constantly over the atomic skeleton. Two mirror-image

compounds of this type cannot be isolated separately because the electrons constantly migrate from bond position to bond position. Such bonds are called **resonance bonds.** The migrating electrons cause opposite charges to be set up in the compound. It takes a large amount of energy to overcome like charges (hence repellent to each other) in putting the compounds together. When these compounds are hydrolyzed (broken apart), a great deal of energy is made available for release. Furthermore, the compounds are unstable, and release energy easily.

Incidentally, the changing of bonds resulting from the attraction set up by charged ions that come near the bonds may explain how certain trace metals act as enzyme cofactors in altering a compound. The trace metal ion charge may pull on a C-double bond-O (C=O) and an H^+ proton may move over from an adjacent site. The bond will be reduced to a single bond (C-OH) by attracting the "extra" bond to another site, such as one between two carbon atoms (which can have from one to four bonds). The compound is altered even though there has been no chemical change in the usual sense of a reaction between two ingredients:

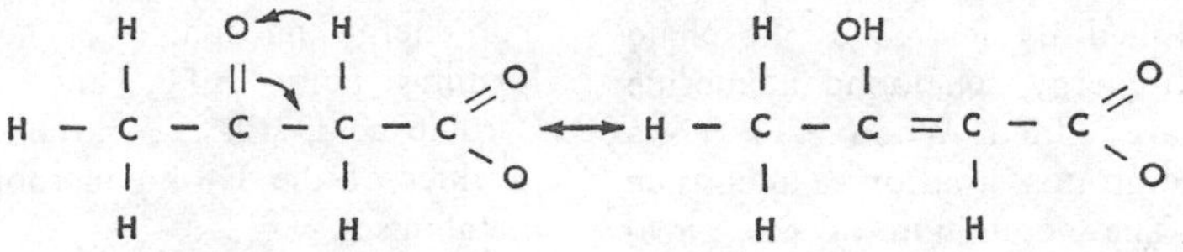

Although there are other energy-rich compounds present, adenosine triphosphate (ATP) is the compound that carries much of the high energy in biological systems. It is composed of a purine ring plus a ribose sugar and may have one, two, or three phosphate groups. See Fig. 93.

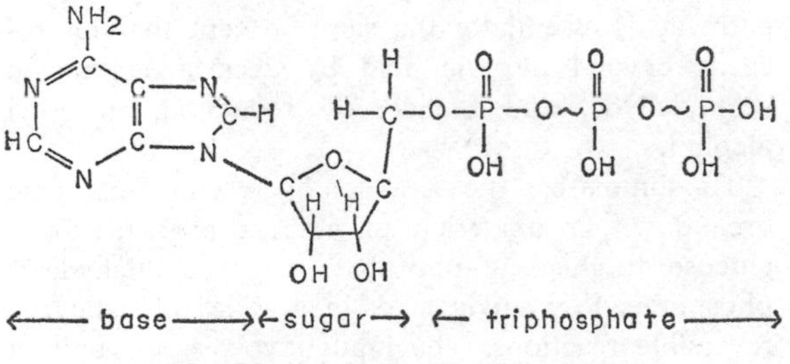

←—— base ——→←– sugar –→ ←– triphosphate ——→

Fig. 93. Adenosine Triphosphate

The monophosphate form has relatively low energy; the diphosphate has increased energy; and the energy of the triphosphate is very high. It can be written

~ Ⓟ to distinguish a high-energy bond from a

low-energy bond—Ⓟ . The higher energy is thought to result from the number of resonance bond locations possible in the triphosphate structure.

The Oxidation-Reduction Mechanism. Before we are ready to follow glucose through its cycle in the cell, there is another principle that we must understand. This is the oxidation-reduction mechanism. Ordinarily, we think of oxidation as the process in which oxygen is added to a molecule of another element. Actually, when oxidation occurs an electron is lost and energy is released. The same objective is accomplished if we remove hydrogen from a compound, instead of adding oxygen.

To remove oxygen from a compound we must put energy into the compound, by adding an electron or by applying heat, for example. This is called **reduction,** and we can accomplish the same result by adding hydrogen to the compound. One process must be a partner of the other.

A compound that furnishes oxygen is an **oxygen donor,** and when it donates oxygen to the compound being oxidized, the donor is reduced. In terms of oxidation by removal of hydrogen, the hydrogen donor is oxidized when it gives up hydrogen; the compound receiving the hydrogen is reduced. In terms of electron transfer, when an electron is removed a compound is oxidized and energy is released; when an electron is added, a compound is reduced and energy must be put into the compound. This can be shown schematically as follows:

1. Oxygen added:

$$A + BO \longrightarrow AO + B$$
$$\text{Recipient} + \text{Donor} \longrightarrow \text{Oxidized} + \text{Reduced}$$

2. Hydrogen removed:

$$AH + B \longrightarrow A + BH$$
$$\text{Donor} + \text{Recipient} \longrightarrow \text{Oxidized} + \text{Reduced}$$

Metabolism of Glucose. The transfer of electrons to produce energy is essentially what is being accomplished by all the chemical activity taking place within the cell. Let us examine the energy pathway of glucose metabolism, called the **Embden-Meyerhof (E-M) pathway.**

The glucose molecule ($C_6H_{12}O_6$) is broken down anaerobically (in the absence of oxygen) into 3-carbon molecules of pyruvic acid ($C_3H_4O_3$). This is done in liver and muscle cells in a series of steps; phosphate groups supplied by inorganic phosphate (PO_4) or by the high-energy compound adenosine triphosphate (ATP) are added. When ATP gives up one of its phosphate groups, it becomes adenosine diphosphate (ADP). The acquisition of the phosphate is called **phosphorylation;** when added anaerobically, it is supplied by the substrate and is substrate phosphorylation; when added aerobically (in the presence of oxygen) it is known as oxidative phosphorylation. This occurs only in the cytochrome system in the mitochondria.

Glucose becomes glucose-6-phosphate in a reaction catalyzed by the enzyme glucokinase, as the first step in the E-M pathway. Glucokinase is helped by the action of insulin, which is secreted by special cells in the pancreas called the islets of Langerhans. (Diabetics, lacking in insulin production, cannot phosphorylate glucose.) Each step in the metabolic pathway requires one or more enzymes and cofactors to catalyze it. Glucose-6-phosphate is a key compound, for it is also formed by fructose and galactose, other hexose (6C) sugars, to prepare them for glycolysis. Glycogen, the animal storage product, is converted for use by its reversion to glucose.

Glucose-6-phosphate is structurally rearranged into fructose-6-phosphate; from this, fructose-1,6-diphosphate is formed by the addition of a second phosphate group supplied by ATP. The numbers refer to carbon atom locations in the sugar structures, which are ring-shaped. Now that we have a 6-carbon, 2-phosphate ring, we can divide it into two parts; it will yield two 3-carbon structures that become glyceraldehyde-3-phosphate (also called phosphoglyceraldehyde or PGAL). Aldolase is one of the enzymes involved; isomerase is another.

The next step involves the anaerobic oxidation of glyceraldehyde-3-phosphate by the removal of hydrogen, which is accepted by NAD, the electron acceptor nicotinamide adenine dinucleotide formed with the vitamin niacin. At the same time, glyceraldehyde-3-phosphate is phosphorylated by the addition of inorganic phosphate (phosphoric acid). This forms 1,3-diphosphoglyceric acid with one phosphate becoming a high-energy bond. ADP, (adenosine diphosphate) now receives the high-energy phosphate, which raises the ADP to ATP, and leaves 3-phosphoglyceric acid (PGA). Three-phosphoglyceric acid is now converted to 2-phosphoglyceric acid by an enzyme, triose mutase.

Dehydration (removal of H_2O) of 2-phosphoglyceric acid is catalyzed by enolase and magnesium ions, forming phosphoenol pyruvate. This is appropriately named PEP, for the dehydration has redistributed the energy in the atomic structure to create a new high-energy phosphate group. The high-energy phosphate is now donated to ADP, forming more ATP, and leaving pyruvic acid. Pyruvic acid, still a 3-carbon structure, is the endproduct of the E-M anaerobic pathway of glucose metabolism.

If anaerobic conditions continue, pyruvic acid will be reduced to lactic acid by donating the hydrogen that was carried off by NAD as NAD·H when the glyceraldehyde-3-phosphate (PGAL) was oxidized. This process occurs in the muscles during exercise. Lactic acid buildup causes sore muscles; when oxygen is supplied, the lactic acid will reconvert to pyruvic acid either for aerobic oxidation or for return to storage glycogen.

In the fermentation of carbohydrates by yeast, the pathway is essentially the same, except that it proceeds beyond pyruvic acid by decarboxylation to acetaldehyde, and thence, by reduction, to ethyl alcohol.

To summarize the E-M pathway, the anaerobic breakdown or oxidation of glucose proceeds from glucose, to glucose-6-phosphate, to glyceraldehyde-3-phosphate, to pyruvic acid, in a series of catalyzed reversible reactions. The input involves one unit of 6-carbon sugar, glucose, plus two NAD, four ADP, two ATP, and two phosphoric acid portions. The output of the pathway gives two units of 3-carbon pyruvic acid, plus two NAD·H, two ADP, four ATP, and two H_2O from dehydration. The energy gain thus is two ATP.

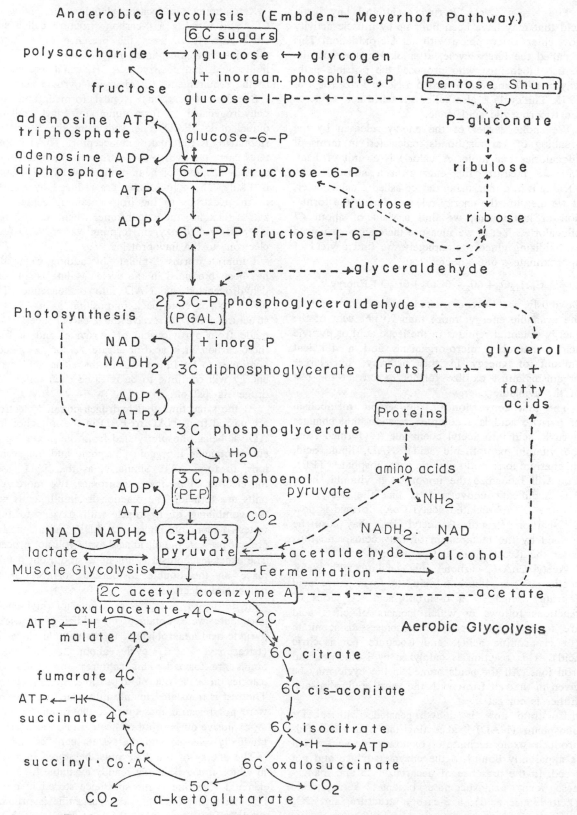

Fig. 94. Anaerobic Glycolysis (E-M Pathway) and Aerobic Glycolysis (Krebs Cycle)

The Krebs Cycle. Pyruvic acid and any lactic acid that may have been built up by muscle activity now enter a new phase with aerobic oxidation. This is called the Krebs cycle, after Sir Hans Krebs, the British biochemist who proposed the possible pathway in 1937 and was awarded a Nobel Prize for it in 1953. The cycle is also called the citric acid cycle or the tricarboxylic acid cycle.

We spoke earlier of the energy released by the breaking of various bonds, calculated in terms of kilocalories per mole. A calorie is a unit of heat, and we measure the energy in a substance by burning it in a measuring device called a calorimeter. If we measure the energy released during the formation of lactic acid, we find a yield of about 47 kilocalories. Yet if we measure the energy produced by oxidizing glucose completely, we find a yield of approximately 686 kilocalories:

$$C_6H_{12}O_6 + 6O_2 \rightarrow 6CO_2 + 6H_2O + \text{Energy}$$

Obviously, we have extracted only a small part of the available energy; more than 90 per cent of the energy potential remains in the lactic acid or pyruvic acid. Anaerobic micro-organisms find a sufficient amount of energy in this pathway, but larger organisms must oxidize further to extract the rest of the available energy.

The first conversion of the aerobic metabolism of pyruvic acid is a complex reaction that changes pyruvic acid into acetyl coenzyme A (formed from the vitamin pantothenic acid), NAD, lipoic acid, magnesium ions, and thiamine pyrophosphate (TPP). You will recognize the important B vitamins here. The significant conversion is that the acetyl coenzyme A (abbreviated acetyl·CoA) formed is now a 2-carbon structure. Acetyl·CoA may also be formed by the metabolism of fatty acids, and thus enter the Krebs cycle at this point.

Acetyl·CoA (2-carbon) now combines with oxaloacetic acid, a 4-carbon acid, to form citric acid (citrate) which is a 6-carbon acid. A series of reactions follows in which isomers of citric acid are formed: an intermediate isomer, cis-aconitate (or cis-aconitic acid), and isocitrate (or isocitric acid). This reaction is catalyzed by isomerase and iron ions. All the acids named in the cycle may be given in the salt form with the word ending in -ate. Either is correct.

Isocitrate now is dehydrogenated, with NAD·phosphate (NADP) accepting the hydrogen. This produces oxalosuccinate or oxalosuccinic acid which is apparently bound to the enzyme surface and not freed. In the presence of decarboxylase and manganese ions, oxalosuccinate becomes ketoglutarate (ketoglutaric acid), a 5-carbon structure, and CO_2 is given off.

Ketoglutarate is decarboxylated again, giving off CO_2 and forming a 4-carbon structure called succinyl·CoA. It requires the presence of cofactors thiamine pyrophosphate, magnesium ions, NAD, lipoic acid, and coenzyme A-SH, which are similar to the requirements for changing pyruvic acid.

Succinate (4-carbon) is next formed, and then dehydrogenated to form fumarate. Flavin adenine dinucleotide (FAD), derived from the B vitamin riboflavin, is the hydrogen acceptor. The flavoproteins carry the hydrogen electrons to the **cytochrome system**—a group of hemoproteins widely distributed in tissue. The cytochromes are reduced by transfer of the electrons to the iron-containing center. Enzymes reoxidize the cytochromes when the hydrogen is transferred to oxygen, forming water. NAD feeds electrons to the flavoproteins.

Fumarate forms L-malate by adding water that has been produced in the cycle. Malate is, in turn, dehydrogenated by NAD into oxaloacetate. This places oxaloacetate back in position as a 4-carbon structure, ready to receive the 2-carbon acetyl·CoA made from pyruvic acid or pyruvate, and to form the 6-carbon citric acid or citrate. As long as glucose continues to be funneled into the citric acid cycle, energy will continue to be released each time a hydrogen is split off.

At the same time that hydrogen-donated electrons are passed from NAD to FAD in the mitochondria, ADP is being phosphorylated from the mixture present, containing inorganic phosphate and magnesium ions, to form ATP. Similarly, as the FAD passes the electron along the cytochromes, two more ATP units are formed. The reaction is simultaneous with the combining of hydrogen with oxygen to form water. If inorganic phosphate, ADP, and magnesium are not present, the citric acid cycle will not proceed. The exact mechanism is not understood, but it takes place on the double membrane surfaces of the mitochondrial cristae.

In terms of energy, two units of high-energy phosphate were obtained in oxidizing glucose to pyruvic acid anaerobically. In the aerobic citric acid (tricarboxylic acid) cycle, about 38 high-energy bonds are formed. This captures about 304 kilocalories in ATP out of a possible 686 in glucose. The rest is translated mainly into heat. If the energy were not released in sequence, the result would be an explosive outpouring of heat. Thus, the energy is gradually released and the system can "capture" it.

The energy of the ATP must be stored in a form in which it can be made readily available for use on demand. Creatine is the substance stored in muscle in most higher animals, while argenine is primarily used in lower animals.

ATP+creatine⇌creatine phosphate+ADP

This reaction releases the ADP to return to the cycle to pick up more phosphate. Creatine phosphate transfers its high-energy phosphate back to ADP to form ATP when it is needed for muscle contraction; it is catalyzed by creatine kinase.

The exact mechanism of muscle contraction is not known. Two proteins, actin and myosin, are present in muscle cell fibrils. Proteins have long, complex molecules that form folded chains or helical spirals, and temporary bonds may form between the **gyres,** or folds, at certain places. During muscle contraction, actin and myosin temporarily unite to form actino-myosin, probably due to the introduction of electron energy donated by ATP. As soon as this energy is used, the proteins separate and contraction stops. When a muscle is stimulated, there is a short latent period while creatine phosphate donates $\sim$ ⓟ to energize the union of actin and myosin. When the energy is used, the two proteins separate and the muscle relaxes. This accounts for the recovery period during which the muscle will not contract, regardless of stimuli, until new energy is received.

When glucose is not available, the body uses up its stored glycogen and then turns to its protein and fat stores. Protein is stored by building tissue, so tissue must be torn down and converted into substances that then enter the general metabolic pathways. Amino acids, the basic building blocks of protein, are 2-carbon structures and may be converted to pyruvic acid or acetate (acetyl·CoA) and enter the Krebs cycle. Fats are 2-carbon structures and also enter the cycle at the acetate level, to be metabolized to CO_2 or recycled. Photosynthesis in plants produces basic 3-carbon units that may be built up to polysaccharides by linking units, or may be metabolized by entering the pathway at the 3-carbon level of 3-phosphoglycerate.

Since pentose sugars such as ribose are necessary to form nucleic acids there is a side pathway often called the **pentose shunt.** There glucose-6-phosphate is transformed to phosphogluconic acid and then to ribulose and ribose. Ribose sugars may be metabolized in the breakdown of nucleic acids by conversion to glyceraldehyde and re-enter the E-M pathway as fructose-6-phosphate.

The building of tissue is an extremely complex operation, with each amino acid having its particular metabolic pathway. It is important to remember that each step requires the presence of a specific enzyme and perhaps other cofactors as well. The lack of one enzyme, or perhaps of some trace metal, can interrupt one step of a process and thus inhibit an entire system. It is amazing that the majority of organisms function normally in view of this complexity.

REPRODUCTION

Methods of Reproduction. The genetic basis of reproduction by means of chromosomal activity has already been discussed. Above the level of the molecular structure of DNA, however, organisms show diverse methods of producing offspring like themselves.

Simple **fission** was probably the original mode of reproduction, either by dividing in half or by dividing into many parts at the same time. Protistans show both methods. The protozoan parasite that causes malaria reproduces inside the red blood cells, then the many offspring burst the cells open, almost like spores are released by plants.

Fission is not restricted to single-celled animals, but is found in a number of multicellular invertebrates. Usually this is not the only reproductive method of the animal, and it may only occur seasonally or in times of stress.

The success of fission as a reproductive method in some multicellular animals may well be related to the high degree of regenerative ability seen in some of the lower invertebrates. Where cells do not differentiate into complex structures and become highly specialized they apparently retain the ability to form the simple structures throughout adult life.

Sponges cut into small pieces will grow back into complete sponges, although they may be somewhat distorted. Planaria, small flatworms found in freshwater streams, will regenerate as whole animals when cut into fragments. Some marine annelid worms (related to garden earthworms) break spontaneously into fragments at certain times of the year to reproduce.

Reproduction by **budding** is a similar mechanism, in which a new animal will grow from some area on the adult. The buds may remain attached to the parent, forming a colony, or they may drop off and begin a solitary existence. In most cases, those animals that bud during part of the year also reproduce sexually at other times.

Sexual reproduction, involving the fusion of gametes to form a fertilized egg or zygote, is the method adopted by the widest variety of organisms and generally by the most advanced groups. Naturally, the premium paid for survival in competition with other groups includes finding the most efficient reproductive system.

Survival. Some means of bringing the gametes (egg and sperm) together to insure fertilization must be found. Many animals solve this problem through quantity: they produce millions of eggs and sperm, releasing them at random into their surroundings. This method is more effective in a marine or aqueous environment where there is no danger of the gametes drying out. There is a high mortality rate on such eggs and sperm, however, for they may drift away in the water without meeting, or may be eaten by other animals.

Protective Coverings. When animals became land dwellers, some protective coverings had to be developed to prevent the eggs from desiccating. The Amphibia (frogs and toads) never quite solved the problem; they still must find a moist environment to lay their eggs, and they surround the eggs with masses of jelly. Reptiles developed a leathery shell for protecting the egg, and the birds improved on this by depositing a calcium shell.

Copulatory Organs. Another adaptation that has developed to insure fertilization is that of copulatory organs. If a shell is deposited around an egg, the sperm must reach the egg inside the female reproductive tract before shell glands secrete the covering. When the egg is to develop inside the female and be born live, insertion of the sperm is certainly necessary. Even when the egg is to be released without shell into the water, many animals have developed some kind of fertilization device. A number of invertebrate animals produce the sperm in packets called **spermatophores.** The packet is then placed inside the female by the male or female, to be dissolved at the proper time. Some of the fish have developed fleshy clasping devices near the anus so they can hold onto the female when she is laying her eggs.

Nutrition. The problem of supplying food to the developing egg, after it has been fertilized, is one that is solved in a variety of ways by various animal groups. Eggs that are to be released into the environment generally contain much yolk, to furnish nutrients for the egg cytoplasm when it begins cell division (cleavage).

Yolk is composed of varying amounts of lipids and glycogen; a fat called lecithin is a prominent ingredient of reptile and bird eggs. Seen in the microscope, the yolk may be in the form of granules or platelets. Phosphoproteins also are found in a number of eggs.

The distribution and amount of yolk in an egg gives us a terminology for the various types of eggs. **Alecithal** eggs are those without much yolk, and are found almost entirely among mammals. **Polylecithal** eggs contain much yolk, as eggs of birds and reptiles. The distribution of the yolk compared with the location of the nucleus and cytoplasm varies also. **Isolecithal** eggs usually have a small amount of yolk evenly distributed throughout the egg. **Telolecithal** eggs have the yolk concentrated heavily at the bottom of the egg opposite the area of the cell nucleus. **Centrolecithal** eggs have the yolk concentrated in the center of the egg with the cytoplasm and nucleus in a layer around the periphery. These are not the only terms in use to describe yolk distribution; many authorities have developed their own terminology. The distribution of the yolk has a profound effect upon the cell division that must take place for the embryo to develop, as we shall see shortly.

Larvae. When a fertilized egg begins to undergo cleavage it soon uses up its supply of yolk for nutrition. A common solution for continuing growth among many animals is for the cleaving eggs to develop into simple larval forms, capable of feeding themselves, as soon as possible. Larval types may range in complexity from simple "gut-tubes" with a little protoplasm and a cilia fringe for motility, to fairly large animals such as tadpoles, the larval stage of frogs. Some parasitic Protozoa and worms have more than one larval stage, to help in transferring them from one host to another. Insects commonly have several larval stages relating to the seasonal availability of food and to temperature. The advantages of having larval stages are obvious, because the eggs cannot contain sufficient nutrients to last until the adult form is capable of feeding itself.

Larvae are usually vulnerable to predators, and because of the large number produced by egg-laying in quantity, they form a good proportion of the food supply for other animals. Many animals feed almost entirely by straining out the tiny organisms, called **plankton,** that float in the sea. This includes, consequently, numerous eggs and larvae.

Retention of the developing egg inside the mother is another device for protecting and nourishing the offspring. This is the most economical method in terms of the number of eggs that must be produced to insure survival of the species.

Egg-laying. Oviparous reproduction is the laying of eggs. A small number of animals, such as certain roundworms and some fish, have developed the system of simply retaining the eggs in the ducts inside the mother after fertilization. There is no connection between the embryos and the mother, but the young are protected and born live. This method is called **ovoviviparity.** Guppies, the tropical fish popular in home aquaria, are ovoviviparous, and the newborn show the remaining egg yolk in the abdomen for a short while after birth. **Viviparous** reproduction produces living young, and the embryos do not depend upon egg yolk for nourishment, but form an attachment with the maternal reproductive tract, called a **placenta,** in order to receive nutrition. The complexity of the reproductive tracts of the various types of animals, both male and female, is related to the type of egg produced, the method of fertilization, and the length of time the egg is to be retained by the female.

Germ Cells. The origin of the germ cells, the cells which form the future gametes, is still a source of disagreement among authorities on development. The future germ cells can be identified in some invertebrates during the early cleavage stages of the developing egg. In higher animals there is evidence that they can be identified in the developing embryo very early, and traced as they migrate into a position beside the primitive kidney. Mesoderm provides accessory cells that surround and support the germ cells, thus forming **gonads**: the ovaries in females and the testes in males.

The gonads are essentially sexless in the developing embryo; the same basic body plan is laid down for both sexes. Apparently hormones produced by the genetic enzyme instructions determine the sex that develops.

Gonads and the Primitive Kidney. In primitive animals the gonads ruptured to release the gametes into the body cavity, from which they were shed into the sea through abdominal pores. Available pores in the lower animals are frequently associated with the excretory system. In vertebrates, the primitive (holonephric) kidney extended almost the full length of the dorsal body cavity wall. The tubules from each segment emptied into a long duct which in turn emptied to the outside through abdominal pores, as in the ancient fish, or into the chamber (cloaca) along with the end of the digestive tube.

The anterior end of each kidney (pronephros) was soon abandoned for the excretory function, and the long gonad lying beside the kidney took over anterior kidney tubules, forming a funnel to transport the gametes to the excretory duct leading to the exterior. The remaining kidney is called the **opisthonephros,** and occurs in some adult primitive fish.

In male vertebrates, the sperm-collecting tubules then take over the head of the opisthonephric kidney. In the female there are no tubules in the ovary, so the kidney ducts degenerate. The female forms a new duct from the body wall, so that the funnel (ostium) leading to it is all that remains of the primitive duct, according to comparative anatomists.

Embryonically in mammals the middle part of the opisthonephric kidney is functional, and is called the **mesonephros.** The anterior tubules in the male form the convoluted collecting tubule, the **epididymis,** which empties into the old excretory duct, called the Wolffian duct or **vas deferens.**

A primitive developmental trend was toward having the gonads share the excretory duct because they had none. Later the trend was toward separating the ducts for gametes from the ducts for excretory products. At the same time a much more efficient kidney evolved from the most caudal area of the opisthonephros. This is the **metanephric** kidney and it forms in the embryo when an entirely new duct, the **ureter,** begins to push outward from

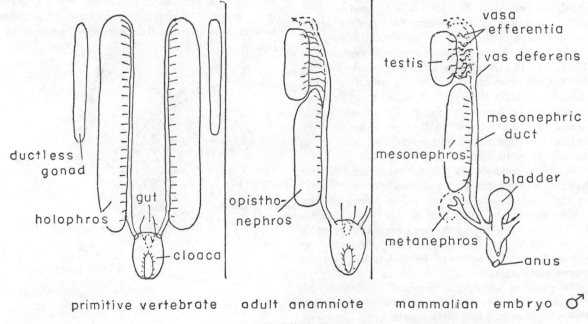

primitive vertebrate   adult anamniote   mammalian embryo ♂

Fig. 95. Kidney Types

the point where the mesonephric duct enters the cloaca. As it grows outward, it branches to form the inner chambers called the pelvis of the new kidney. It is surrounded by tissue from the old caudal opisthonephric area.

MAMMALIAN REPRODUCTION

During the embryological development of mammals the germ cells become arranged in cords of cell clusters in both males and females. These clusters lie along the genital ridge of the mesonephric kidney and are covered by the same tissue that covers the kidney. The germ cells, which will form the gametes in the adult, become part of the tissue covering the gonad when it separates from the kidney. The system is sexless at this point, and two separate sets of ducts arise, one of which will develop and the other degenerate when sexual determination takes place. Exactly what triggers the development of the sex that is contained in the chromosomal recipe is not known. Determination is apparently both enzymatic and positional; in other words, the tissues must be in the right place at the right time to receive enzymes, produced perhaps by neighboring tissue.

The Female Organs. When a female develops, the gonads become paired ovaries, and the germ cells lie in an epithelium covering them. Connective tissue fills in the bulk of the ovary, containing interstitial and secretory cells. The female ducts (oviducts) may have originally been formed by a split in the opisthonephric duct but this is not clearly understood. In primitive vertebrates the paired oviducts emptied separately into the urogenital sinus, but gradually the ducts began to fuse into one duct at the sinus. The fusion progressed upward, forming the **vaginal chamber** and then the **uterus,** enabling animals to retain the egg in the uterus for development. Various vertebrates show different amounts of fusion in the ducts, so that in some animals the vagina and uterus are still partitioned and young develop in both sides of the uterus.

In other animals the ducts have fused at the terminal end but the uterus divides into two "horns" as the ducts separate to go to the paired ovaries. Young develop in both horns of this type of uterus. In humans, the uterus is somewhat triangular in shape, but in developing, the embryo retraces the paired-duct stage, followed by fusion to form the uterus and vagina.

To produce an egg, a germinal epithelium cell divides mitotically, and the cell (oögonium) thus formed moves into the interior (cortex) of the ovary. See Fig. 97. Here it undergoes first meiosis and is

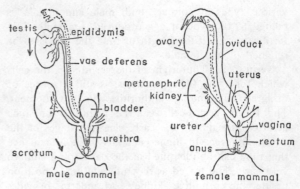

Fig. 96. Urogenital Systems

surrounded by secretory cells to form a **primary follicle.** Some primary follicles are present in the ovary when an infant is born. The follicle cells continue to accumulate until they are several layers deep, forming a translucent zone, the **zona pellucida,** around the egg. Connective tissue forms a covering around the area, enclosing a fluid-filled cavity and the egg with its follicle cells. It is now called a **Graafian follicle.** Regnier de Graaf (1641–1673), a Dutch anatomist, first observed the follicle and thought he had discovered a mammalian egg. It was not until 1827 that the embryologist Karl von Baer found the tiny ovum inside the follicle.

The follicle moves out to the surface of the ovary as it develops, and, when ripe, the ovum bursts from the follicle with its liquid and is swept into the ostium (funnel) of the oviduct. The follicle becomes a bright yellow body (the **corpus luteum)** that secretes hormones. Meanwhile the egg travels down the oviduct covered by cells called the **corona radiata.** If it is fertilized it undergoes the second meiosis.

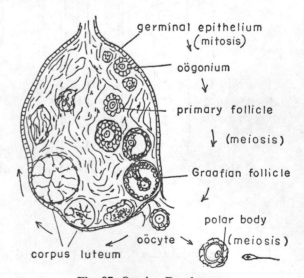

Fig. 97. Ovarian Development

The uterus lining, the **endometrium,** is thick, highly vascular, and filled with crevices. A fertilized egg burrows into this tissue and erodes the surrounding tissue to establish contact with the maternal blood supply.

The Male Organs. When the male system develops, the gonad undergoes changes similar to those occurring in the female at first. The germinal epithelium forms into cords which lengthen into extensive tubules. However, the tubules then move to the interior and become covered with a tunic of connective tissue. The germinal epithelium forms the inner lining of the tubules along with interstitial cells. The germ cells divide mitotically to form spermatogonia. These cells then undergo two meiotic divisions to become primary and secondary **spermatocytes** as they move into the lumen of the tubule. The resulting cells, called **spermatids,** become imbedded in special interstitial cells called **Sertoli cells.** These cells "nurse" the spermatids until they are mature, growing the long tails that give them mobility. When ready for release in the tubule, the sperm head contains little more than the nucleus, the Golgi apparatus, centrioles, and mitochondria.

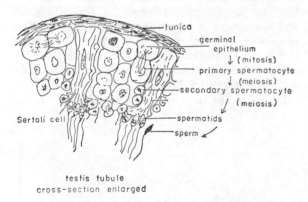

Fig. 98. Enlarged Cross-Section of Testis Tubule

Many sperm collect in tubules, called **vasa efferentia,** and enter the much-coiled epididymis. They then descend to the **vas deferens,** the abandoned mesonephric duct of the kidney that functioned during embryological development.

The paired ducts enter a single urogenital duct (urethra) just below the area where the ureters enter from the adult metanephric kidney, at the base of the bladder (see Fig. 96). The ducts develop accessory seminal vesicles at the caudal ends, where secretions are formed to accompany the sperm through the ejaculatory duct section and into the urethra. Around the urethra lie Cowper's glands and the prostate gland, both of which contribute to the fluid called **semen.**

Embryonic Sexlessness. During embryological development the external structures also show sexlessness. The urogenital sinus and anus (the digestive tract opening) are separated by a ridge. On either side of the urogenital sinus lie a pair of folds, the genital folds, and lateral to these folds are rounded genital swellings. Ventrally, in the center, is the genital tubercle.

If the individual is to become a male, the genital tubercle lengthens into the **penis,** the upper part of which contains spongy tissue that may in the adult become engorged with blood to stiffen it for mating (copulation). During embryonic growth the urethra that empties the bladder exists only as a groove extending from the original urogenital sinus opening along the underside of the genital tubercle. The groove and sinus close gradually until they are incorporated into the penis as the urethra. The genital folds become the prepuce ensheathing the penis, and the genital swellings alongside become scrotal pouches that will contain the testes.

If the individual is to become a female, the genital tubercle becomes the **clitoris,** a very small mound of tissue lying in the center anterior to the labia minora (small lips) that are formed from the genital folds on either side of the urogenital sinus. The sinus becomes divided into the urethral (urinary) opening and the vaginal opening leading to the uterus. The genital swellings form the labia majora (larger lips) that lie on either side of these structures.

Descent of Testes. The gonads of both sexes arise along the genital ridge of the mesonephric kidney. How then do the testes come to lie on the scrotal pouches outside the abdominal cavity? The answer is that both ovaries and testes descend somewhat from their original location into the body cavity as their weight increases and the ligaments suspending them are stretched. The testes, however, continue a downward migration, probably due to the differences in the growth rates of the body and of the ligaments suspending the testes from above and pulling down on them from below. Sperm are incapable of developing fertility if they are subjected to warmth such as that developed by the warm-blooded birds and mammals. Therefore, the testes migrate to the scrotal pouches, where surface cooling maintains a lower temperature than is present internally. In some animals the testes descend to the scrotal pouches seasonally, and are retracted the remainder of the time. It is thought that the custom in some cultures of sitting in hot baths for long periods of time arose as a birth control device.

The reproductive systems in humans do not become functional until the early teens, when cyclic activity of the systems becomes apparent, especially in the female. In some animals the male shows sexual activity only at certain seasons, but man and his relatives, the other primates, are functional during all seasons of adult life.

The Estrus Cycle. Female animals show periods of heightened sexual activity and willingness to mate alternating with periods of inactivity. This periodic change is called the **estrus cycle,** and the mating period is known to animal breeders as "heat." In the primates and man, a more obvious occurrence in the cycle is menstruation, characterized by the discharge of blood and cellular debris from the uterus at about four-week intervals. For many years the correlation between the estrus and menstrual cycles was not understood, and it was thought that ovulation took place during the menstrual cycle. It is now clear that both estrus and menstruation are separate events of a single cycle, involving a complex interplay between various organs in the body, adjusted to insure mating at the optimum period for fertilization of the ovum.

Female Hormones. The **pituitary gland** is a small tag of tissue that protrudes from the undersurface of the midbrain. It secretes a number of hormones, substances that are released into the bloodstream and carried to other parts of the body where they stimulate various activities. At the beginning of the female cycle the pituitary secretes the follicle-stimulating hormone (FSH). This acts upon the ovary, stimulating the germ cells to divide and form a primary follicle, which will mature and become a Graafian follicle.

As the primary follicle develops, it begins to secrete the estrogen hormone estradiol, which acts upon the lining of the uterus and vagina. The uterine lining proliferates, becoming thicker and producing mucus in preparation for the arrival of a newly fertilized ovum. Estrogen also affects the behavior of the female, encouraging mating at the time of ovulation. As the level of estrogen in the blood increases, it also affects the pituitary, signaling it to decrease FSH production and begin production of the lutenizing hormone (LuH). The maturing Graafian follicle, having migrated to the ovary surface, erupts, releasing the ovum surrounded by the zona pellucida and additional cells that form the corona radiata. What remains of the Graafian follicle is now called the corpus luteum and is stimulated by the LuH to produce another hormone, progesterone. Progesterone causes the uterine lining to secrete, and increases the blood supply, preparing the uterus to receive the ovum.

The progesterone level in the blood stimulates the pituitary to begin secreting a lactogenic hormone (LaH) at the end of the phase, causing the mammary glands (breasts) to swell and the milk ducts to develop. If the ovum is not fertilized during this time, the corpus luteum begins to degenerate; with the decrease in progesterone, the breasts regress and the uterine lining begins to break down. The blood flow into the area is reduced and the pooled blood and fluid already in the wall tissue is released, along with the degenerating mucous membrane. This forms the menstrual flow.

If fertilization does take place, the corpus luteum

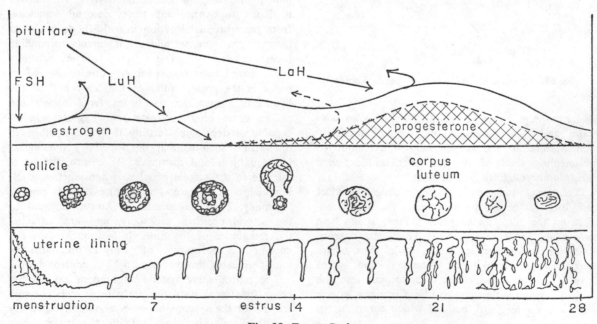

Fig. 99. Estrus Cycle

continues to secrete progesterone to maintain the uterine lining during pregnancy. At the same time, progesterone suppresses the production of FSH by the pituitary to prevent further ovulation during a pregnancy already in progress. It is this factor that is the basis of many birth control pills. Progesterone taken orally in conjunction with other substances simulates the conditions present in pregnancy. These conditions cannot be maintained indefinitely, and menstruation must be allowed to occur periodically by stopping the medication. If this is not done, in a few months the pituitary will override the progesterone hormone and ovulation will occur in spite of the oral hormone.

Progesterone medication may also be used to re-establish the cycles of individuals suffering from irregularity, and to overcome low fertility. When menstruation is suppressed for several months, the cessation of medication apparently initiates considerable overflow of FSH from the pituitary, and multiple ovulation occurs. A high incidence of twins and other multiple births has been recorded in such cases.

The menstrual cycle usually repeats itself every twenty-eight days, and ovulation (estrus) is believed to occur most generally between the thirteenth and fifteenth days of the cycle. About seven days elapse between the time of fertilization and the implantation of the developing ovum.

Fertilization. It has been thought that sperm survive in the female tract for only one or two days after insemination, but this is open to question. Sperm retain mobility for a longer period than they retain actual fertilization capacity. The irregularities that occur in the female cycle and in the length of time sperm retain viability make birth control by means of the so-called rhythm method very unreliable.

Apparently the number of sperm introduced also affects the chance of fertilization. Although sperm are motile, the chances of a given sperm surviving a migration through the vagina and uterus and into the oviduct are small. Muscular contractions of the uterus and oviducts may assist the passage of the sperm. The pH of the reproductive tract may have a destructive effect on the sperm as well. Only one sperm usually fertilizes an ovum, but in many animals a number of sperm heads may become partially embedded in the ovum. It is possible that enzyme secretions of many tiny sperm are needed to assist one sperm to penetrate the coating of the ovum.

Ova are believed to secrete two substances called **gynogamones;** one causes sperm motility to be stimulated and the other causes sperm heads to become sticky so that they will adhere to the egg surface.

Sperm are thought to secrete two **androgamones.** The first is to suppress activity in order to conserve the small energy supply in the sperm until it is stimulated by secretions in the urethra and by the gynogamones. The second androgamone apparently aids in ovum penetration. Hyaluronidase is the enzyme secreted by the Golgi apparatus in the sperm head. It dissolves hyaluronic acid that helps to hold the cells of the corona radiata around the ovum.

Once a sperm head penetrates the egg, the nucleus of the ovum undergoes its final meiotic division, extruding the last polar body. The male and female pronuclei then unite, usually by aligning themselves along a mitotic spindle to undergo first cell division or cleavage.

Fertilization acts as a stimulus to the egg, causing a number of changes in the cytoplasm. Other stimuli can induce at least limited development in some kinds of ova; such development without the male nucleus is called **parthenogenesis.** A pin prick, temperature shock, or chemical trauma may cause parthenogenesis.

There has been much speculation as to how the sperm activates the egg. Perhaps it introduces mitochondria to produce energy for synthesis, or enzymes that unblock the messenger RNA to begin protein synthesis in the ribosomes.

The sperm entry usually causes cortical granules from the egg to join the membrane surrounding the egg, and a fertilization membrane thus is formed. The egg seems to shrink away from this membrane and no other sperm can penetrate to the egg.

Pigmented eggs often show a streaming or slipping of the surface cortical material, exposing other internal areas. The site of the sperm entry may determine the plane in which cell division (cleavage) will first occur, or the location of the future invagination, called the **blastopore,** which occurs during development.

It is obvious that the single fertilized ovum, or zygote, must contain the potential to direct the formation and growth of an entire organism. It has been shown that different areas of the cytoplasm of the undivided egg already contain different potentialities. The egg has polarity; the top of a floating egg is the animal pole, where the cytoplasm containing the nucleus lies. The opposite end is the vegetal pole, which contains more of the yolk, and the area that will become the gastrula (gut). If the egg is cut in half at the equator, the animal pole will produce ectodermal and sensory cells, but no gastrula. The vegetal pole will produce a gastrula, but no neural tube. The substances from both poles must balance each other to produce a normal embryo.

The planes of cleavage of the fertilized egg, plus

the differences in the cytoplasm in various areas, will soon cause dividing cells to differ in potentiality from one another. As the ovum divides, the fate of various cells produced must inevitably be determined so that the descendants of a given cell will produce a certain tissue or organ while other cells become assigned to other structures.

Cleavage. Once an egg is fertilized, its pattern of development depends upon the species involved. There are basic patterns which are characteristic of several phyla, and variations of these patterns which are found only in a small number of animals in a single species.

The two basic patterns of cleavage are called **radial** and **spiral** cleavage. Spiral cleavage occurs in the great majority of the invertebrate phyla from lowly flatworms through the more advanced annelid worms, mollusks, such as clams and snails, and arthropods, which include animals such as insects, shrimps, and crabs.

Radial cleavage probably arose as a modification of the more ancient spiral pattern, and is found chiefly in the echinoderms, such as starfish, and in the chordates, which include all the vertebrates or animals with backbones.

The first two cleavages of the ovum usually occur in the vertical plane at right angles to one another, to produce four cells. It is at this point that differences between spiral and radial cleavage begin to be seen.

In spiral cleavage, the third cleavage occurs in the horizontal plane, at right angles to both previous divisions. Since most invertebrate eggs are moderately yolky, the third cleavage occurs more toward the animal pole, making the lower four cells larger than the upper ones. The mitotic spindles occur at an oblique angle, so the upper cells (**micromeres**) are rotated and lie to the left or right of the lower cells (**macromeres**) like a stack of cannon balls seen from above. See Fig. 100, top.

In 1892 an American embryologist, E. B. Wilson, worked out the details of spiral cleavage and developed a system of numbering the cells (**blastomeres**) through subsequent divisions that makes it possible to trace their fate. After the first division produces a quartet of micromeres and macromeres, the macromeres divide again, placing a new quartet of micromeres, 2a, b, c, d, between the original micromeres and the four macromeres. The macromeres are now labeled 2A, B, C, D. A third cleavage and a fourth take place, producing a third quartet of micromeres, 3a, b, c, d, and a fourth quartet of micromeres, 4a, b, c, d. The macromeres are now 4A, B, C, D. Meanwhile the first and second quartet have continued to divide so that the **blastula,** a solid

or hollow ball of cells, depending on the species, is now formed. See Fig. 102.

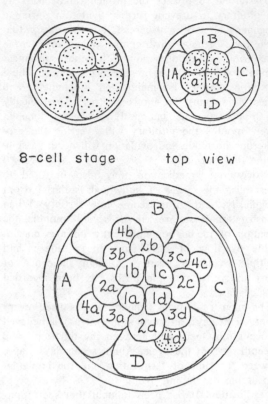

Fig. 100. Spiral Cleavage

The first three quartets of cells and their offspring cells will form the ectodermal germ layer which produces the animal covering plus nervous and sensory structures. In the lower invertebrates ectoderm also forms ectomesoderm which will provide whatever mesenchyme and limited muscle strands occur. Lower invertebrates do not have true mesodermal muscle layers or bundles. The fourth quartet of micromeres plus the four macromeres push into the interior to become endoderm and form the digestive tract. The endoderm cells may also contribute to mesenchyme and muscle strands as endomesoderm. This stage is called a gastrula and the invaginating process is gastrulation. In higher invertebrates one micromere, 4d, forms all of the mesoderm that furnishes the true muscle layers, plus gonads and mesenchyme.

There are, of course, numerous variations to this general pattern but the fate of these cells is determined at the first cleavage. If the first four blastomeres are separated mechanically and cultured, only a quarter of an embryo will be formed from each cell. If cell groups later are transplanted from one area of the embryo to another, they will still form the predetermined structure or area. This is known as **determinate cleavage.**

Animal groups sharing similar developmental patterns are considered to be more closely related to each other than to those with differing cleavage patterns, even though their adult morphology may be quite different. They may have had a common ancestor. The phyla showing spiral, determinate cleavage are said to be members of the Annelid-Arthropod line.

Radial cleavage differs from spiral cleavage in that the mitoses always occur at right angles to the cell plane, drawn through the animal-vegetal poles, instead of at oblique angles. This produces cells stacked directly one above the other. Since the cells are not as compactly arranged as they are in spiral cleavage, they soon form a hollow ball of cells, the blastula. This type of cleavage is seen primarily in the echinoderms and chordates, plus a few rather obscure groups. Here again variations occur among different species. See Fig. 101.

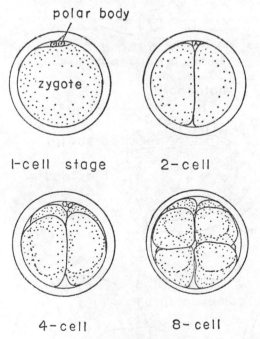

Fig. 101. Radial Cleavage

If the first four cells undergoing radial cleavage are separated mechanically and cultured, each cell has the ability to form a complete embryo. This is known as **indeterminate cleavage.** In these groups it is possible to transplant cells at much later stages and have them adapt to their new surrounding in-

stead of producing the structure they normally would. The stage at which a cell's fate becomes fixed varies among the organs and structures.

Another distinction that shows the divergence of the two lines is seen in the fate of the area known as the blastopore. When the round blastula invaginates to form the digestive tract and become a gastrula, it is as though a finger had been poked into a child's soft rubber ball. A sac is formed within a sac, and the area where the indentation occurs is known as the blastopore. In spiral, determinate cleavage lines, the blastopore area eventually becomes the mouth of the animal. These are called **protostome animals.** In radial, indeterminate cleavage lines, the area of the blastopore becomes the anus, and the mouth forms at the opposite end. These animals are called **deuterostomes.**

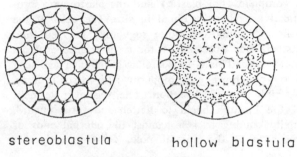

Fig. 102. Blastula Types

In some of the lowest invertebrate groups, a solid blastula (stereoblastula) is formed. The internal digestive layer may form by splitting off cells from the outer layer, or by in-wandering of cells from one pole, or from all sides. This is a mesenchymal type of formation, where cells move more individually than where they move as a sheet in the more usual invagination of the hollow blastula. See Fig. 103. The solid blastula is probably the more primitive mode of formation. Digestion is within each cell (intracellular) rather than in a common digestive tube where enzymes are secreted into the tube (extracellular).

Little or no growth occurs during cleavage. The amount of cytoplasm stays about the same, but the nuclear material has been repeatedly duplicated by cell division.

GASTRULATION

It is possible, by tagging certain blastula cells with vital stains, to follow their activity when invagination takes place to form the gastrula. It is not gastrulation itself that determines what the various cells become. This was determined earlier during cleavage. Those cells at the vegetal pole of the blastula migrate into the interior through the blastopore to form endoderm (or entoderm), the digestive tract germ layer. The animal pole cells become the ectoderm, the germ layer that produces the nervous system and skin surface of the embryo. Cells in the marginal or equatorial zone of the egg will form the rodlike notochord found in all chordate animals at some part of their life cycle, and also the mesoderm that forms the muscles, body wall lining, and kidney.

The pattern of gastrulation in the chordates depends greatly on the amount of yolk and its distribution in the cells. *Amphioxus,* one of the simplest chordates, is a fish-shaped lancelet. The yolk of an *Amphioxus* egg is evenly distributed so that cleavage is complete (holoblastic) and the blastomeres produced are virtually equal in size. The cells of the hollow blastula are quite regularly arranged, and gastrulation proceeds without interference from heavy yolk. This is also similar to the cleavage seen in the nonchordate echinoderms such as sea urchins.

Amphioxus Gastrulation. The vegetal pole cells of the blastula begin to flatten and push inward, finally coming to rest against the internal ends of the cells at the animal pole, forming a double-

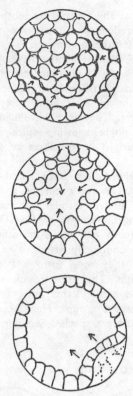

Fig. 103. Gastrulation Methods. *Top to Bottom—* Splitting, In-wandering, Invagination

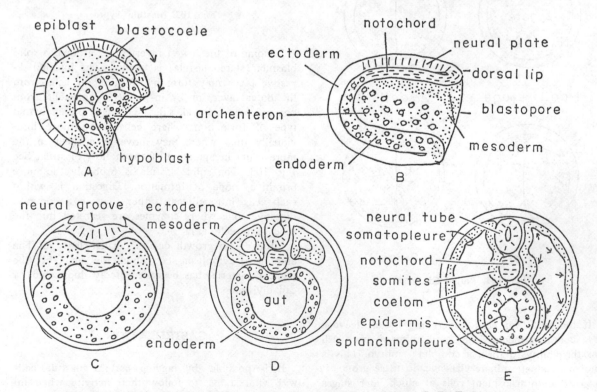

Fig. 104. Amphioxus Gastrulation. A, B–Longitudinal Stereo-sections. C, D, E–Cross-sections

walled cup. The embryo then begins to lengthen as the opening narrows. The cavity of the blastula, the **blastocoele,** has been eliminated, and the new cavity that is formed is called the **archenteron** or primitive gut. The opening of the archenteron is the blastopore. Cells migrate from the area around the outside of the blastopore, passing over the rim or lips and turning to the inside when the archenteron is formed. Because this cavity is the primitive digestive tract, the stage formed is called a gastrula.

The mechanism that initiates gastrulation is not known. In some ova it is known that the vegetal pole area is softer than the surrounding surface cells. The cells near the animal pole have a faster growth rate, with a much higher oxygen intake, than the vegetal cells. Perhaps the pressure exerted by these rapidly growing cells causes the vegetal cells to move inward where there is less resistance. This is not the whole story, however, because when gastrula cells are separated by the enzyme trypsin and allowed to reaggregate, the vegetal pole cells go to the interior while the animal pole cells attempt to arrange themselves on the surface.

The archenteron tissue of *Amphioxus* contains the future endoderm cells ventrally and laterally that will form the digestive tract. Dorsally and dorsolaterally, however, are areas destined to form, respectively, the notochord and the mesoderm. The dorsal lip of the blastopore seems to be the site of greatest activity, with the presumptive notochord cells turning inside first, as the endoderm turns in at the sides and bottom.

The presence of notochord cells in the dorsal roof of the archenteron induces changes to take place in the exterior cells directly above. Tissue culture experiments indicate that secretions from notochordal cells cause the ectoderm to thicken, forming the **neural plate.** The neural plate drops beneath the ectoderm which closes over above it. The plate forms a groove and then closes to become the neural tube, and eventually the brain and nervous system. Notochord transplanted experimentally beneath other ectodermal areas that do not normally form neural plates will induce neural tube formation in the new area.

It is interesting to note here that the first experiment in tissue culture was reported by the American biologist Ross G. Harrison in 1907. He aseptically excised some frog neural tissue and grew it in a hanging drop of frog lymph on slides. His pioneering work showed clearly a method for studying living cells. The Nobel Prize committee recommended him for an award in 1917, but it was not conferred. Again in 1933 his work was considered, but was rejected on the grounds that the value of tissue culture was limited and the work was old. In 1954, three persons were awarded the Nobel Prize for studies on tissue culture of viruses on embryonic tissues. Much of the current research on diseases and in embryology is performed using tissue culture. Harrison died in 1959, unrecognized.

The notochordal cells in the dorsal region of the *Amphioxus* archenteron form into a bulge that extends the length of the roof; this rounds up into a tubular bundle and is pinched off from the archenteron, forming the rodlike notochord. At the same time the notochord is forming, similar bulges occur in the archenteron on either side of the notochordal area. Each detaches from the gut to form a tubular bundle of mesoderm lying beside the notochord in the space between endoderm and ectoderm.

The mesodermal bundles grow down into the space between the gut and ectoderm, the tissues migrating along those structures and using them as a substrate. Fluids secreted into the center of the bundle help to force the growing tissue back against the two surfaces. Thus the body cavity becomes lined with mesoderm, forming a true coelomic cavity. The tissue meets "back to back" above and below the gut and becomes supportive sheets for the organs. These sheets are called **mesenterics.**

The mesoderm that overlies the gut endoderm forms the digestive musculature; together the two layers are called the **splanchnopleure.** The mesoderm underlying the ectoderm forms the dermal layer of the skin and musculature; these two layers together are called the **somatopleure.** The mesoderm lying along the notochord and neural tube forms muscle segments called **myotomes.**

Only when the prospective mesoderm and endoderm have become located inside the embryo during gastrulation can the surface outside be correctly called ectoderm. In order to avoid giving the impression that mesoderm and endoderm are formed from ectoderm, some authorities refer to the exterior surface as the epiblast (or protoderm) and the vegetal cells as the hypoblast. The actual germ layers cannot be considered formed until inward migration of cells has been completed.

The process of gastrulation differs between species and is altered in pattern by the physical barriers of excessive yolk. The process of evolution, while carrying similar patterns from one group to another, has probably led to short-cutting or blurring over of some of the steps that took place in the ancestral groups. Amphibians such as frogs, toads, and newts show the effects of moderately large amounts of yolk, while reptiles and birds show very large amounts. Mammals often exhibit the short-cutting or blurring over steps seen in the lower vertebrates.

Amphibian Gastrulation. In amphibian eggs, cleavage occurs more rapidly in the animal pole cells than it does in the cells at the vegetal pole, so that the animal pole has many small cells while the vegetal pole has fewer, larger cells. The blastocoele cavity is displaced upward and the blastopore begins to form on the side of the blastula above the heaviest yolk area.

The site of the blastopore in frogs can be predicted at the time of fertilization of the ovum. Cortical slippage of the pigmented outer layer, under the stimulus of fertilization, exposes an underlying area of gray yolk known as the **gray crescent.** It is the crescent area that becomes the blastopore site, showing clearly that areas of differentiation exist prior to cleavage.

Cells migrate at gastrulation from all over the blastula surface, a process known as epiboly, and turn inward at the blastopore lips (involution). In 1938 the German embryologist Hans Spemann identified the dorsal lip as the most active site, calling it the primary organizer. This area of the embryo contains the potential to direct development of an entire embryo if transplanted experimentally to another area at the early stage of gastrulation. As gastrulation continues, different cells pass over the dorsal lip area, and only partial induction can be obtained with that tissue experimentally.

Because of the yolky cells at the vegetal pole in amphibians, a uniform invagination such as was seen in *Amphioxus* cannot occur. Cells begin to move inward at the dorsal lip, forming the archenteron, primarily composed of future endoderm. At first some of the dorsal archenteron cells are prospective notochord and mesoderm, sometimes called the chordomesoderm. As the archenteron pushes back into the blastocoele, however, the chordomesoderm is separated from the roof. This eliminates the steps of outfolding and detaching from the endoderm that were seen in *Amphioxus*.

The blastopore opening gradually becomes round, but the center is filled with a plug of yolky cells. As the yolk is used by the embryo this barrier decreases and cells are able to migrate into the interior from all around the lips of the blastopore.

Presumptive mesoderm cells that will migrate to the interior form a band around the original blastula surface not far from the future blastopore site. When the blastopore opens, the future endoderm moves to the inside. The notochord mesoderm then moves in at the dorsal lip and mesodermal sheets move in along either side of the notochord, which pushes forward above the archenteron toward the future head region. Eventually these sheets split in the center, forming the coelomic cavity, and the mesoderm covers the gut and body wall as it did in *Amphioxus.*

Another German embryologist, W. Vogt, in 1929 succeeded in developing **fate maps** of the gastrula. These show exactly where and how the primary germ layers develop in certain amphibians.

Amphibian species show a number of variations in their patterns of development. One is the way in which the neural tube forms. In most vertebrates, including man, the neural tube is formed by infolding. First a thickened longitudinal area of cells, the neural plate, forms. Then a longitudinal groove in the center of the plate forms as the sides of the plate rise up into the folds. The folds come together and fuse, closing the groove into a tube which then splits away from the surface ectodermal epithelium. Some amphibians, however, show a different method, in which a solid longitudinal thickening forms. The thickening then hollows out internally to form a tube. In some amphibians different regions of the neural tube will show different methods of origin even in the same animal.

Gastrulation in Fish and Birds. The effect on cleavage and gastrulation of eggs with massive amounts of yolk is seen in some fish and more so in birds. Cleavage occurs only in a small disc of protoplasm on the surface of the egg, and no division of the yolk takes place. This type of discoidal cleavage is said to be **meroblastic** or incomplete as compared with **holoblastic** or complete cleavage.

The cleaved disc or blastoderm also cannot gastrulate and form a blastopore in the manner seen in the frog, because cleaved cells do not surround the yolk. In some fish, the edge of the blastodisc acts as a dorsal lip and cells migrate under the posterior edge to form the entoderm lying over the surface of the yolk.

The mesoderm follows the entoderm, converging along the midline and pushing dorsally. Eventually the blastodisc grows sufficiently to cover all of the

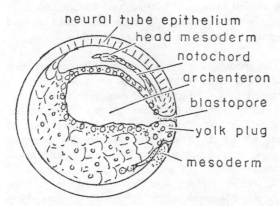

neural tube epithelium
head mesoderm
notochord
archenteron
blastopore
yolk plug
mesoderm

Fig. 105. Amphibian Gastrulation

unused yolk except for a plug area. It is in the region dorsal to the area where the yolk plug finally appears that the endoderm first turns to the interior.

In birds gastrulation has become even more specialized, being limited to a small area in the center of the blastodisc. The proliferation of cells in the blastodisc causes it to lift off the yolk in the center. This translucent area is called the **area pellucida,** while the outlying portion of the blastodisc resting on the yolk is the **area opaca** (opaque area).

At the future caudal end of the chick embryo, cells thicken and begin to migrate from the raised blastodisc down toward the yolk to form the hypoblast, or future endoderm. There is disagreement as to whether the future endoderm splits off as the lower layer of blastoderm cells or whether there is actually an inward migration of some of the surface (epiblast) cells. Perhaps both processes occur.

The Primitive Streak. The thickened caudal surface area changes from a wide triangular area into a lengthened concentration of cells called the **primitive streak.** A groove is formed and cells begin to migrate toward it, turning inward all along the groove. At the head of the groove is a depression known as the **primitive pit.** Here cells migrating from the anterior surface area form a hump known as **Hensen's node,** where the cells turn inside and migrate back beneath the area from which they came. This anterior tissue forms the notochord and head mesoderm.

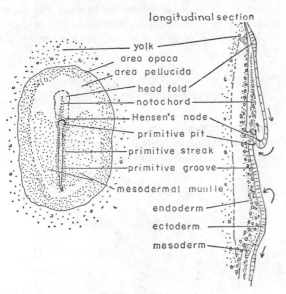

Fig. 106. Primitive Streak Stage in the Chick

Again the mobility of the cells at gastrulation is seen, as those from the lateral areas of the blastodisc move in and turn under along the primitive groove, forming ridges at the edges. These migrating cells also move back beneath the surface, forming the mesodermal mantle on either side of the streak. This extends forward on either side of the notochord as the latter pushes forward beneath the surface.

The rodlike notochord, by pushing straight forward beneath the surface, causes the surface to become elevated to form the head fold. The mesoderm mantle does not fill in under the head fold, and it is in this area that the ectoderm will later break through to the endoderm to form the mouth.

The extra-embryonic tissue, surrounding the embryo proper, eventually grows down over the yolk so that the hypoblast encloses the yolk in a yolk sac. Apparently the original hypoblast does not contribute much of the actual gut, which is thought to be formed principally from the endoderm cells that migrate from the area of the primitive streak. The migration of cells at the primitive streak justifies considering it as a specialized blastopore.

Mammalian Cleavage and Gastrulation. Birds and mammals are thought to have evolved from different lines of reptiles. The reptiles have a yolky egg with a leathery shell; birds have a yolky egg with a calcareous shell. Mammalian embryology shows its inheritance of developmental patterns of the vertebrates that preceded them in evolutionary history. For example, the mammalian egg will be nourished by attachment to the mother, and so very little yolk is stored in the egg. Yet a yolk sac, which contains nothing at all, forms during development. Other processes seen in birds or reptiles have been modified or eliminated by the specialized conditions of internal development.

Mammalian eggs undergo the first cleavage stages while descending the oviduct after fertilization. The mitoses are not synchronized, as they are in most invertebrates and the lower vertebrates, but occur one at a time so that three-, four-, five-, and six-blastomere stages are found, and so on up.

Some of the lower mammals form a hollow blastula with a blastocoele cavity, but most mammals form instead a solid ball of cells called a **morula,** or mulberry stage. Sooner or later, however, the cells sort themselves into an inner cell mass and an outer enveloping layer. This stage is called a blastocyst because it was formerly thought that this outer layer of cells did not enter into the formation of the germ layers, serving only as a protective capsule. This layer will form the **trophoblast** or nourishing layer when it becomes attached. See Fig. 107.

The inner cell mass separates from the outer layer except at the top. The cell mass was thought to split off or proliferate the hypoblast cells to form the yolk sac. There is now some evidence that the outer layer forms the hypoblast by cell division at the point of juncture with the lower edge of the

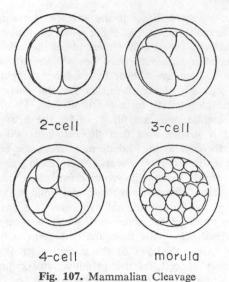

Fig. 107. Mammalian Cleavage

inner cell mass. It is quite possible that hypoblast cells may come from both areas.

The inner cell mass next splits above the center, forming a cavity by secreting fluid into the split to expand the space, creating the amnionic membrane and cavity. The remaining cell mass below the cavity constitutes the epiblast and here a short primitive streak will form later, initiating gastrulation. See Fig. 108.

In the mammal, the mesoderm cells apparently begin to form first by proliferating from the trophoblast wall. These cells line the trophoblast and migrate to the yolk sac, covering it. They also push between the amnionic membrane and the trophoblast where they are in contact. Mesoderm forms all of the blood vessels and corpuscles, and in this way the formation of a blood supply for the embryo is hastened. The primitive streak forms later in mammals than it does in birds, so that mesoderm from

that source would not organize quickly enough to supply nourishment.

As in birds and reptiles, the cells along the primitive streak migrate inward and down and then back under the surface. Some of the loose cells seem to migrate down toward the hypoblast to help form endoderm. The cells that migrate between the epiblast and hypoblast form the mesoderm of the embryo proper.

The Notochord. Those cells that turn under anteriorly at Hensen's node form a compact process that is the rudiment of the notochord, seen so prominently in the chick. The notochord process is rounded, and in some birds and mammals, including man, is tubular. This canal is thought to be homologous with the archenteron of primitive vertebrates, and is called the **archenteric canal.** Different species vary in the degree of canal formation. In some, the curved rod fuses with the sides of the hypoblast below in the manner of the primitive archenteron. Later the two structures separate again and the notochord forms a tube in the manner reminiscent of *Amphioxus.*

The mammalian primitive streak recedes toward the posterior as the notochord pushes forward, seeming to stretch the embryo. When the notochord is excised experimentally, the embryo fails to lengthen normally. The notochord degenerates in higher vertebrates during development, as it becomes surrounded by the vertebrae. It contributes to the intervertebral pads, the cushions between the vertebrae. These can become herniated (bulge out), producing the painful back injury known as the "slipped disc."

PROTECTIVE MEMBRANES

It is necessary to examine the protective membranes developed in birds in order to compare them with those formed in mammals. The change from

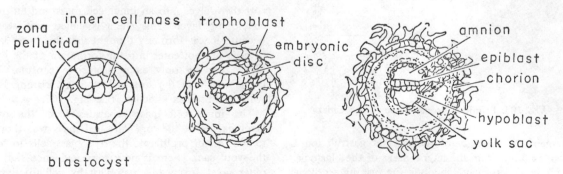

Fig. 108. Mammalian Gastrulation

self-contained nourishment to that of a maternal supply has changed the extra-embryonic membrane somewhat, but the relationships can still be seen.

At the completion of the primitive streak stage in chicks the head is raised up off the surface of the expanding area outside the embryo. In similar fashion the growth of the notochord and neural tube caudally causes the tail fold to raise up.

The gut is a closed tube only in the head and tail region. It opens onto the yolk sac to the gut by the **yolk stalk.** A diverticulum of the gut in the tail region is the **allantois,** which forms the bladder in the adult animal. The yolk stalk and allantois are enclosed and will become narrowed to form the umbilical cord.

Chick Amnion Formation. The ectoderm and mesoderm (somatopleure) begin to raise up in a fold around the embryo proper. Gradually the folds increase until they fuse above the embryo, forming a closed chamber around it. Fluid secreted into the chamber protects the embryo from friction and jarring. The layers separate after the folds fuse, leaving the membrane called the amnion, composed of ectoderm and mesoderm, surrounding the embryo. The outer membrane, composed of mesoderm and ectoderm, is called the **chorion** or serosa.

As the mesoderm continues to grow, it eventually lines all of the outer membrane and covers the yolk sac and allantois. Mesoderm forms all of the blood supply of the animal. Clusters of cells form blood islands, and these in turn form tubules, with the cells inside becoming the blood corpuscles. The tubules grow together (anastomose) to form the circulatory system. Thus the embryo becomes supplied with membranes that are highly vascular. Nourishment from the yolk is picked up by the blood supply of the yolk sac.

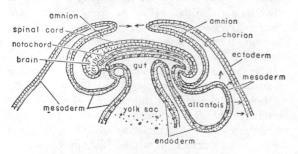

Fig. 109. Extra-embryonic Membranes

The chick allantois diverticulum of the gut grows rapidly into the space between the embryo and the chorion. The excretory wastes are deposited in the allantois in crystalline form as uric acid, for storage. In chicks, a layer of protein albumin, the egg white, surrounds the embryo and its membranes; this, like the yolk, is absorbed for additional nutrition. The growing allantois expands to lie against the chorion, enabling the blood to absorb oxygen that comes through the shell into the air space in the egg. Some authorities prefer to reserve the term "chorion" for the mammalian membrane and use "serosa" for the external extra-embryonic membranes in chicks. See Fig. 109.

Mammalian Membrane Formation. In different mammals the amnion is formed in various ways, intergrading from the method seen in chicks to the one previously discussed in which the inner cell mass splits. Rabbits and carnivores show an intermediate condition. The outer cell layer, or trophoblast, disappears directly above the inner cell mass. The entire inner cell mass forms the embryonic disc, which is now exposed. The extra-embryonic somatopleure folds up over the embryo in the same way that it does in chicks. In other mammals, such as guinea pigs, hedgehogs, and anthropoids (apes and man), these steps are eliminated and the amnion is formed by a splitting of the inner cell mass, so that the embryo is never exposed.

The size and degree of vascularization of the yolk sac and allantois vary also. In some mammals the yolk sac is large; in others the allantois is the principal membrane that performs nutrition exchange. In humans, a small yolk sac and a small allantois form. These become invested with mesoderm that proliferates from the trophoblast wall; mesoderm from the same source also lines the trophoblast itself. The trophoblast in humans appears to be able to erode the uterine lining to burrow into it.

The trophoblast becomes highly villiform, increasing surface contact with the uterine epithelium. The allantois never becomes large, nor does the yolk sac, but the allantoic mesodermal covering proliferates greatly, forming a highly vascular membrane that fuses with the trophoblast to line a part of the extra-embryonic cavity. These fused membranes, the chorion, form the placenta; the umbilical cord contains the yolk sac remnant and the allantois stalk, with its prominent blood vessels that supply the fetus. The membranes still serve the purposes of protection and nutrition, as they did in the chick. In the mammal, however, nutrition is accomplished by diffusion or active transport from the mother to the fetus through the placenta.

As the fetus grows, only one area of the membranes, the placenta, remains attached to the uterine wall; the fetus, inside amnion and chorion, grows out into and eventually fills the uterine cavity.

DETERMINATION

The fantastic amount of differentiation that must take place in moving from zygote to blastula to gastrula and thence to adult form is almost beyond comprehension. Determination and guidance of the sequence of steps must be directed by the DNA of each cell.

Some **inducer substances** are apparently protein enzymes that are exported to surrounding cells. Other inducer substances may be lipids, nucleo-proteins, or steroids. The level of these substances in the cell will "shut off" production of that particular inducer in one cell or cell group and "turn on" production in adjacent cells. Adjacent cells may produce substances that will in turn pass instructions to other cells while they themselves are forming a particular structure.

The cells receiving an inducer substance must have a potential to react to that inducer or they will not perform the task being directed. During certain limited times, many cells have the potential to evolve into more than one structure even though they would normally become one given structure or part of a structure.

We said earlier that the notochord seems to stimulate neural tube formation, and other structures or cell groups appear to perform similar inductive functions. It is apparent that interactions are extremely important in normal development. Much experimental work is being done in this field with the use of transplants, biochemical analysis, radioactive tracing, and other techniques to unravel the secrets.

Differentiation. The gastrula stage provides for the formation of three primary germ layers in all of the higher invertebrates and in all the vertebrate animals. In turn, these embryonic germ layers subdivide to form specialized tissues and organs. The tissues and organs then develop into the adult organ systems. We can represent this development by means of a diagram (Fig. 110), but it must be remembered that all systems are not at the same level of development at the same time.

Induction is a continuing process during the entire period of differentiation. It provides for initiating, continuing, or stopping the various activities. It will be many years, if ever, before scientists are able to identify all of the complex interacting substances involved. The basic modes of action of the cells and tissues are nevertheless typical, regardless of the organ system in which they occur.

Epithelial Tissue Behavior. Epithelial tissues, the sheetlike tissues formed primarily by ectoderm and endoderm, and less frequently by mesoderm, behave in certain distinctive ways. The cells are blocklike and sticky; if separated artificially, they will adhere to one another and re-form sheets of tissue.

One of the ways in which epithelia may form structures is by **thickening.** This may occur either when cells in a certain area become columnar or when cells from adjacent areas migrate into a given position. Neural plate, for example, may be formed in either of these ways.

Thickening is usually the first step taken toward several other types of formations. **Folding,** a major developmental activity, will often follow thickening action. Linear folding may produce long tubes such as the neural tube. Infolding or invagination of a small, more or less circular area will form hollow vesicles such as the skin glands. In general, ectoderm tends to fold inward, while endoderm evaginates (folds outward).

Splitting is another major mechanism in differentiation. While we have seen that tubes may be formed by folding of thickened areas, they may also be formed by internal splitting of similar thickened areas. Some vertebrates form neural tube one way and some the other; in some amphibians, one region of the tube follows one mechanism while another region forms with the other, even in the same animal.

When splitting occurs, fluid is secreted into the split, causing the formation of a cavity. Mesoderm formed during gastrulation may be aided in lining the body wall and in covering the gut by the secretion of fluid into a split in the center of the tissue. The space so formed is the coelomic cavity.

Clefts, formed perpendicular to the surface of bundles of tissue, also may be seen. Mesoderm bundles split in this way produce segmentation of the bundles, or masses. This forms the blocks of tissue called somites that occur along the caudal region on either side of the developing neural tube.

Splitting is also the method whereby a linear tube detaches from the surface that formed it by folding. The neural tube, when closed over, must detach itself from the ectoderm that lies above it.

Outgrowths, or **diverticula,** may form a number of structures. We have mentioned the villi of the trophoblast (chorion) of the human embryo when it becomes embedded in the uterine lining. These villi are outgrowths of the surface that help to increase the exposed surface area and aid in holding it against another surface. The internal surface of the developing intestine also forms numerous villi. Gill filaments in fish are formed by similar outgrowths that increase respiratory surface. The villi may be hollow when formed, or they may be solid at first and then split. Once hollowed, blood vessels and

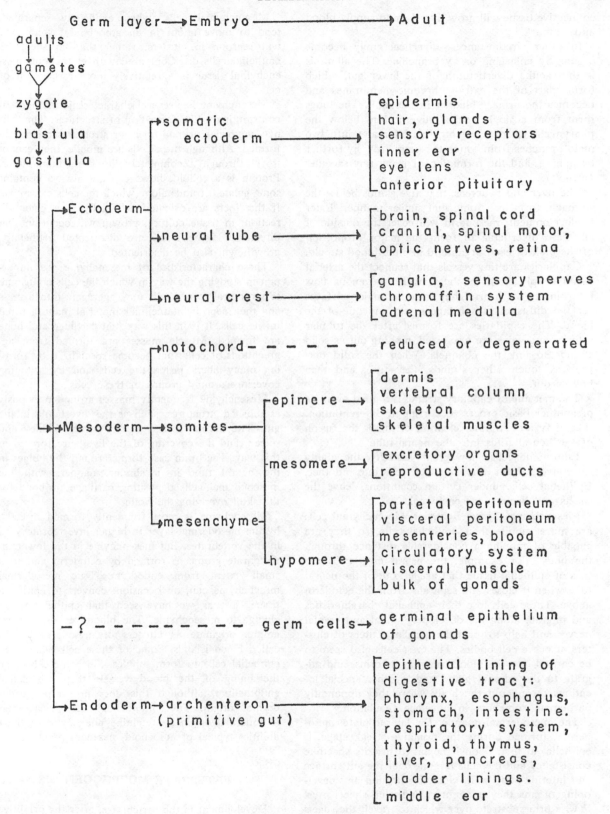

Fig. 110. Differentiation

connective tissue will grow into them to supply blood and support.

In other circumstances, diverticula may become organs by enlarging or by branching. The allantois is the saclike diverticulum of the lower gut, which forms part of the extra-embryonic membranes and becomes the urinary bladder in the adult. The lungs form from a single ventral diverticulum below the pharyngeal (throat) region; this bifurcates into two main branches from which the lungs grow by further branching, and the formation of many tiny saccules (alveoli).

The liver and pancreas, digestive glands below the stomach, form in a somewhat similar manner. Later in development, mesoderm will cover the outside of all these diverticula to form the major proportion of the bulk of each organ and also the blood supply.

Capillaries are tiny vessels that connect the arterial blood supply from the heart with the venous flow returning to the heart. They form the major surfaces for the diffusion of substances into and out of the blood. The capillaries are formed after the tubular arteries and veins are organized, pushing out as solid strands to link the channels. When the solid outgrowths touch other strands, they fuse, and then hollow out.

The merging of capillary strands is a good example of another basic process occurring in differentiation, that of **fusion.** Another example of this is the fusion of the neural folds into the neural tube.

Embryonic cells show great **plasticity,** the ability to convert from one type of cell into another. Epithelial cells, under certain conditions, leave the sheets and become mesenchymal.

Mesenchymal Tissue Behavior. Mesenchymal cells are more mobile than epithelial cells, so they are capable of moving a considerable distance through the body. For example, the neural crest cells are tags of epithelial tissue that are left out of the neural tube when it closes and separates from the ectoderm above. These cells lose their epithelial characteristics and migrate along the developing spinal and cranial nerves and help to form the **ganglia,** centers of clusters of nerve cell bodies. The crest cells also seem to be carried into the bloodstream as it forms, and migrate to the adrenal gland and to other special locations throughout the body where they apparently participate somehow in hormone regulation.

The mesodermal mantle (or lateral plate mesoderm), formed during the primitive streak stage, is epithelial at first. Soon it begins to lose its sheetlike consistency and the cells migrate over the gut surface and interior body wall, clustering around any developing organs they encounter. Epithelial sheets must have a firm substrate over which to travel; the sheets cannot move through space or liquid.

Mesenchymal cells, when artificially separated, tend to move about in an amoeboid fashion and will regroup in clusters, rather than in sheets as epithelial cells do. Compared with epithelium, mesenchymal tissue is a relatively loose association of cells.

Mesenchyme has several characteristic methods of contributing to the formation of structures. The cells may **aggregate** around fibers or structures previously formed. Although these cells are mobile, they cannot travel through a completely liquid medium either. Protein is a colloid, however, and always contains some gelated strands along which the cells can move. If the fibers are oriented artificially in a given direction, in tissue culture, growth will occur in that direction. If the fibers are disoriented artificially, growth will also be disoriented.

These characteristics of mesenchyme are important in studying the way in which the cells contribute to structural form. Cells may aggregate into a mass and then deposit intercellular material manufactured in the cells. It is in this way that cartilage and bone are formed. Muscle masses are formed from aggregations of cells that become specialized by forming many fibers inside the cells and by secreting coverings around groups of these cells.

Mesenchyme frequently masses around previously established structures, such as the diverticula in the gut. The fibrous tissues that cover the kidney and spleen, and the covering of the liver, are formed in this way. The brain case forms around the bulges in the neural tube in a similar manner, with the mesenchymal cells depositing cartilage or bone for the skull overlying the brain.

Mesenchyme is most frequently formed directly, by the mesodermal layer in higher invertebrates and in the vertebrates. But mesenchyme in the lower invertebrate groups is formed by ectoderm and to a small extent from endoderm. True mesodermal mesenchyme can, on occasion, convert to epithelial tissue, just as we have seen that epithelium can change to mesenchyme. The blood vessels, for example, originate as clusters of mesodermal cells, called blood islands. Some of these cells change to epithelial cells to form tubules. Since these become the lining of the blood vessels they are called endothelium, although this does not imply endodermal origin. The endothelial cells reveal their mesodermal origin by retaining phagocytic (eating) abilities typical of amoeboid mesenchyme cells.

VERTEBRATE MORPHOGENESIS

Development in the vertebrates, after the primitive streak stage, is remarkably similar among the various animal groups. The embryos are roughly spherical until the notochord begins to stretch the body into an elongated form and produces the "tail bud." The brain and spinal cord begin to develop specialized areas of the neural tube immediately.

After the body is stretched somewhat, it divides into head and trunk regions. All vertebrates develop visceral clefts, openings from the pharyngeal region of the gut tube leading to the outside. In adult fish, these are the gill slits, and they are supported by cartilage on either side of the openings. Amphibian larva show visceral clefts also, and all of the vertebrate embryos possess them at some stage during development. The number and development of visceral clefts is reduced in reptiles, birds, and mammals, leaving room for a "neck" region on the animal.

Limb buds produce changes in embryonic shape, first appearing as humps of tissue, then growing out into paddle-like protuberances. Snakes, of course, have no legs, but traces of this development can be seen.

The heart bulges out from the body during the early stages of development. As the embryo grows, the heart becomes enclosed and this then divides the trunk region into chest and abdomen. At the same time the body connection to the extra-embryonic membranes becomes restricted to the umbilical stalk. These changes are common to all the vertebrates.

In the first eight weeks of development, human embryos form all of the major organs and assume a "humanoid" form. After this time, the primary activity of the fetus is one growth, although the details of structure continue to be refined. For this reason any trauma to the embryo during the first eight to twelve weeks will produce severe or lethal deformities.

Defective Development. Virus infections in the mother during the first months of pregnancy, such as German measles, will almost always produce defective offspring. A number of drugs, if taken by the mother, can also damage the embryo. Ether administered to the mother during the first month may seriously disrupt closure of the neural tube, for example. The difficulty in controlling these problems is that severe damage is often done in the first month, before the mother is aware that she is pregnant.

Premature birth also places a severe strain on the developing fetus, even though the major period of organogenesis is past. The lungs are not adequately expanded and the secretions that help to eliminate mucus from them are not present until the tenth month (280 days or full term). Also the kidneys are not able to handle water balance and excretion

adequately earlier. A fetus born prior to the seventh month is considered incapable of surviving; the closer to full term that birth occurs, the better the chances of survival. Even at full-term birth the infant has not completed development. The reproductive tracts, for example, do not complete their development until about the twentieth year. Overall growth also continues until that time, and it is said that the nose never stops growing.

DEVELOPMENTAL ANATOMY

The study of anatomy in adult animals, without consideration of the nature of development, often neglects the most significant aspects. By examining the major organ systems as dynamic, developing systems, we better understand both form and function. For this reason we will briefly examine the systems in the course of attaining their adult form.

THE NERVOUS SYSTEM

The first structure to begin morphogenetic activity (to assume organic or structural form) is the neural tube. The period in which the neural tube is developing is sometimes called the **neurula stage.** When the tube is in the process of fusing along the dorsal edges the anterior portion closes first, leaving a hole, the **neuropore,** at the anterior tip of the tube. Caudally, the tube is slow to fuse, leaving an open area, the sinus rhomboidalis, lying just in front of the retreating primitive streak.

The closing hollow tube assumes a regularly segmented form, consisting of eleven divisions or neuromeres; in a chick embryo this has developed at about twenty-four hours. Very quickly, however, the anterior three segments fuse and enlarge into a rounded structure, the forebrain or **prosencephalon** (cephalon is the head). Segments 4 and 5 fuse to form a smaller midbrain or **mesencephalon.** The remaining segments, 6 through 11, although losing some of their definition, remain relatively unchanged as the hindbrain or **rhombencephalon.** This is known as the **three-vesicle stage.**

By thirty hours in chicks, the prosencephalon has developed two large lateral swellings, called the optic vesicles, which will later become the eyes. The human brain reaches a somewhat similar form about three weeks after fertilization, but with a much larger forebrain that is slower to close.

The forepart of the brain extends beyond the notochord rod and begins to bend around its tip. At the same time, at about thirty-eight hours in chicks, the anterior portion of the embryo begins to turn on its side. The embryonic connection to the yolk is pushed back into the midsection as tissue folds in under the head. The embryo continues to bend (flex) until it forms a reverse question-mark shape, lying on its left side. Later, when the tail becomes raised off the yolk, the embryo will assume a C-shape. The mammal, since it is not attached to a yolk mass, lies in the C-shape form very early.

The prosencephalon, mesencephalon, and rhombencephalon constitute the **brain stem.** These are the three basic compartments of the vertebrate brain and provide the fundamental nerve pathways for brain function.

During development, the three subdivisions are very soon transformed into five. The prosencephalon divides into the forward telencephalon and the diencephalon. The mesencephalon stays about the same, and the rhombencephalon divides into the metencephalon and myelencephalon.

The diencephalon area is curved around the end of the notochord. Characteristically it has a ventral outpouching, the **hypophysis,** that actually is in contact with an outpouching of the pharynx **(Rathke's pouch)** for a short time. The hypophysis incorporates part of Rathke's pouch to form the pituitary gland.

The dorsal surface of the diencephalon in the lower vertebrates has often developed one or more outpouchings. These may include the paraphysis, epiphysis, parietal body, or pineal eye. The last may actually have a lens, or it may be only a light-sensitive organ.

In the lower vertebrates the optic vesicles move out to the surface, leaving a stalk connecting each eye to the brain at the area of the optic lobes. These lobes tend to move back and above the mesencephalon region in the lower vertebrates. In mammals the visual perception centers are in the new area, the cerebrum, that appears above the telencephalon, but the reflex centers remain above the mesencephalon.

When the human brain has developed embryonically to the five-vesicle stage, it is comparable to the adult stage of many of the lower vertebrates.

In fishes, reptiles, and the lowest mammals such as the opossum, the telencephalon is primarily a center for smell (olfactory lobes). In all of these animals the gray nerve cell bodies occupy the inner wall of the brain and the white myelin sheathing of the axons (nerve cell fibers) is on the outside.

In the mammals (and possibly in a few of the reptiles that were ancestral to the mammalian line) the gray matter began to migrate to the outside of the telencephalic region. This allowed a tremendous expansion to take place in the number and mass of nerve cells, and the cerebral hemispheres began to be developed. It is in this area that the ability to think and to reason is centered.

As the mammal line evolved, increases in the size of the cerebral hemispheres became obvious. Folding of the surface further increased the cell capacity, as seen in the cat brain and, of course, in man. The cerebral hemispheres form most of the mass of the entire brain in man. Only a tiny area remains of the olfactory lobes.

The diencephalon became thick-walled, containing the main fiber tracts connecting the cerebrum to the other parts of the nervous system. The roof (dorsal surface) of the diencephalon remained thin and covered with a network of vascular tissue (the tela choroidea) that permitted the exchange of nutritive materials and wastes between the blood supply and the fluid inside the brain and spinal cord. Another such area is the dorsal surface of the myelencephalon.

The mesencephalon remained small, but in man it bears four small bodies dorsally. The anterior pair are for visual reflexes; the posterior pair are for visual and auditory reflexes.

The metencephalon expanded dorsally to become the **cerebellum.** This is the center of complex muscular activity, and is highly folded in very active animals like the cat. The structure looks like a pair of pine cones.

The myelencephalon became the **medulla,** a thickened area at the base of the brain leading into the spinal cord. Most of the cranial nerves originate here.

Nerve Transmissions. To understand the purposes of the separate areas or structures of the brain it is necessary to examine the principles involved in the transmission of nerve impulses.

The **neuron** is the only cell capable of transmitting a nerve impulse, which seems to be an electrochemical "message." The nerve impulse may travel in any direction within a single cell. A change in the permeability of the cell membrane allows the influx of sodium ions into the nerve fibers and an outflow of potassium ions at the location of the impulse. As the impulse passes along the fiber, the ions revert to their normal concentrations. In the resting state the nerve fiber is surrounded externally with a positively charged field, while inside there is a negative charge. As an impulse passes, the influx of sodium ions changes the internal charge to a positive one, and the impulse thus becomes a wave of positive charges traveling down the fiber.

Some structural arrangement of the cell membrane must be responsible for pumping excess sodium out of the cell at the appropriate time. The mechanism was called a **sodium pump** for some years, although no clear idea was held as to the nature of the pump. Recent research has indicated that round vesicles lying between the double layers of the cell membrane are capable of turning around in place. Ions would flow into the vesicle on the outside and then the vesicle would turn around, dumping its contents inside the cell. This is still very much in the stage of conjecture, however.

The nerve impulse travels slower than an electrical current, but the speed varies according to the nerve structure. Impulses in myelinated (sheathed) nerves are faster than in nonmyelinated nerves. The thicker the sheath, the faster the transmission. It is thought by some that the impulse jumps between nodes of myelinated nerves to move faster. Invertebrates do not have myelinated nerves and the speed is in proportion to the thickness of the nerve. Regardless of whether the stimulus is sensory or motor, nerve transmissions are all electrochemically alike.

The simplest animals to have a nervous system are the coelenterates, which include the jellyfish and *Hydra*. There, the nerve cells form a nerve net, and transmission occurs in all directions from any stimulated point. The nerve cell is either stimulated directly or makes contact with a sensory cell by way of one protoplasmic "arm" and with a muscle cell, or effector, via another protoplasmic "arm" or fiber. This system is suitable only for very simple, sedentary, or slow-moving animals.

Some of the coelenterates began to develop a more specialized mode of action which has been widely adopted by higher groups of animals. Here, one neuron acts as a sensory transmitter and another acts as the motor messenger. This means the message must be transferred between cells across a gap, called a **synapse,** because the neurons do not touch each other. A fluid, acetylcholine, is secreted at the axon tip to furnish the pathway at the proper time. This is perhaps done by the axon tip or by the **glial cells** that are located around the axon. An "eraser" enzyme, cholinesterase, is secreted at the dendrite tip to remove the acetylcholine and end the transmission of each stimulus. Transmission occurs in one direction because the nerve impulse or message can only stimulate the proper sequence of secretions in one direction. The impulse passes through the cell from dendrite to axon, across the synapse, and to the next dendrite.

The next step in increasing specialization is the introduction of an intermediate neuron, called an **association neuron,** between sensory and motor neurons. By the use of multiple branches, it is possible for a stimulus to be directed to the most appropriate effector, or to select the amount of response by the number of effectors stimulated. Since a given muscle cell can only respond in an "all or nothing" manner, this is very important as a means of facili-

tating the degree of action necessary to reply to a given stimulus. This system is seen in insects and vertebrates, which represent the highest developed groups of animals.

Nerve Cell Differentiation. In our study of the development of vertebrates we have seen the formation of the neural tube and the elaboration of brain structures. Within the neural tube and in the neural crest the cells undergo active reproduction and migration; this allows them to set up reflex arcs and association neuron patterns.

In the neural tube, the columnar cells that formed the neural plate begin to differentiate into spongioblasts and neuroblasts. Ciliated ependymal cells form which will line the interior canal, circulating the cerebrospinal fluid. The spongioblasts also move out from the canal lining into the center of the tube wall where they act as germinal cells and by mitosis give off the glial cells, while neuroblasts give off neurons. Some authorities think that spongioblasts may also change into neuroblasts. See Fig. 111.

The cell bodies of the neurons remain, for the most part, inside the marginal layer of the spinal cord. Their axons may grow out the ventrolateral portion of the cord to form segmentally arranged motor nerves. Other neurons remain entirely within the spinal cord and become association neurons linking the various areas of the cord and brain longitudinally. It is believed that once a neuron has been produced by mitosis of a neuroblast, it is incapable of reproducing itself.

The spongioblasts first produce glial cells called **astrocytes,** either fibrous or protoplasmic. These cells have many perivascular feet with which they clasp the blood capillaries that invade the area, and they carry nutritive materials to the neurons. Since some neuron axons and dendrites become very long, the cell body could not possibly maintain the entire neuron.

Later, other glial cells are produced by the spongioblasts. The oligodendroglia ("other-branched glia") seem to aid in forming the coverings that sheath some nerve fibers. The **choroidea** (capillary tissues) over the diencephalon and myelencephalon contribute the microglia cell, a phagocytic, amoeboid cell that "cleans" the neurons. In tissue cultures photographed with time-lapse methods, the microglial cells appear to scamper over the axons, devouring debris. They are the clown cells of the tissue culture film world. The spinal cord then becomes covered with a mantle layer in which the cells have no nuclei.

Neural Crest Cells. While the cells are differentiating inside the neural tube, those from the neural crest area lose their epithelial unity and break up into segmentally arranged tags. The tags form a

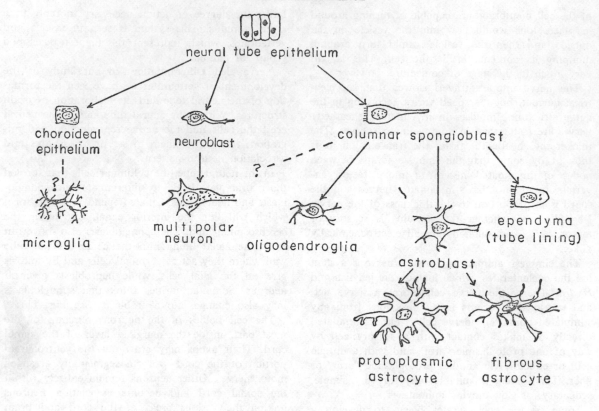

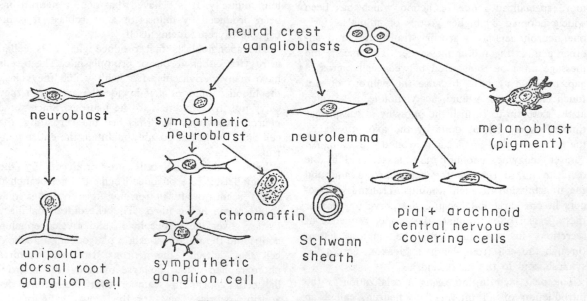

Fig. 111. Cells of the Nervous System

dorsolateral extension or root directly above the location where the motor axons grow out ventrolaterally. The crest cells, called **ganglioblasts,** divide by mitosis and migrate down the dorsal root to meet the ventral extension.

Ganglioblasts can also form neuroblasts, and some of these settle on the dorsal root. These cell bodies establish the dorsal root ganglia, and their axons grow into the dorsal area of the spinal cord to form sensory neurons.

The segmented dorsal and ventral roots meet lateral to the cord, merge, and then immediately branch. One main branch (ramus; *pl.* rami) leads to the skin, one leads to the skeletal muscles, and one passes internally to the visceral organs controlled by that particular spinal nerve. Reflex arcs are set up along the spinal column at each spinal nerve. The simple reflex arc functions as follows:

A stimulus at the skin passes through the dendrite of the sensory neuron to the cell body in the dorsal ganglion. It then passes from the dorsal ganglion into the interior gray matter of the spinal cord, along the axon. The terminal fibers at the end of the axon synapse with the multiple dendritic ends of a motor neuron cell body in the gray matter or with an association neuron that will in turn synapse with a motor neuron. The stimulus passes out of the spinal cord through the ventral root on its way to a muscle cell, which will react to the sensory stimulus.

Crest cell neuroblasts migrate farther down the ventral rami and establish a chain of eighteen pairs of **sympathetic (autonomic) ganglia** lying along either side of the spinal cord in the chest and abdominal (thoraco-lumbar) region. In forming the sympathetic ganglia, the neuroblasts become neurons whose dendrites synapse with the motor axons from the spinal cord.

Fibers (axons) lead from the sympathetic ganglia in a number of paths. Some post-ganglionic fibers pass back to the upper rami leading to skin and muscle and some pass to adjacent sympathetic ganglia. Some of the axons from the spinal cord pass on through the sympathetic ganglia and extend to ganglia formed farther away in the region of the organs innervated.

Other neural crest cells become specialized to form the **pial** and **arachnoid** cells, which form the covering of the spinal cord. The neurilemma cells that secrete the fatty myelin coat of some nerve fibers may also be formed from crest cells.

There are actually few direct reflex arcs that do not involve association neurons within the spinal cord. The association neuron may transfer the nerve impulses on the same side, carry them diagonally across the gray matter, or carry them up or down

the length of the cord. It is this lengthwise transport that involves the various brain sections in the co-ordination of reflexes and in selecting responses to stimuli.

Specialized tracts leading to the brain can be seen in cross-sections of the spinal cord. Bundles of sensory fibers are located primarily in the dorsal part of the cord, while bundles of motor fibers are in the lateral and ventral areas.

Medulla. The medulla or myelencephalon is the reflex center for controlling many of the nonvoluntary activities of the internal organs. A cluster of nerve cell bodies, called **Dieter's nucleus,** completes reflex arcs for respiration, heart rate, blood pressure, swallowing, and vomiting. The tenth cranial nerve, which greatly affects the viscera, is centered here, as are several other cranial nerves.

Cerebellum. Nerve fibers that pass ahead through the medullary area may synapse in the cerebellum, the metencephalon region, at the reflex centers called the **dentate nucleus** and the **deep cerebellar nucleus.** In man, some of the cell bodies have moved to the outside of the cerebellum, above the myelinated nerve fibers. **Purkinjie cells** are found in the outer layer. These cells are unique to this area and have intricately branched dendrites. Muscle co-ordination is centered in the cerebellum, and a bridge of fibers has formed across the brain on the ventral side to co-ordinate both sides of the body. Damage to the cerebellum does not produce paralysis but causes a loss of co-ordination.

Midbrain. The base of the midbrain or mesencephalon contains the sensory tracts that have passed up the spinal cord and under the cerebellum area on their way to higher centers, plus the descending motor fibers and those leading from the cerebrum back to the cerebellar area. The roof of the midbrain develops the superior (anterior) paired **colliculi** and the inferior (posterior) paired colliculi. The former are associated with the nuclei of the third cranial nerve; the latter with the fourth cranial nerve. Both nerves affect eye muscles. The **red nucleus** is in the basal tract area. The large number of tracts passing through the midbrain cause the walls to be thickened until the central cavity is much reduced. This forms the **aqueduct of Sylvius** and connects the cavity of the cerebellar area (metacoele or fourth ventricle) with the cavity of the diencephalon (diocoele or third ventricle).

Diencephalon. The diencephalon can be likened to a switchboard: all the impulses to the thought centers in the cerebrum must pass through it. Association neurons dominate the basal region, or **hypothalamus,** and none of their axons leave the central nervous system. A series of nuclei associated with autonomic function are present. These regulate body temperature, water balance, appetite, carbohydrate and fat metabolism, blood pressure, and sleep. They are co-ordinated here with the medulla reflex centers. Up to this level, all actions directed by the brain are automatic, not requiring thought.

The sides of the diencephalon, the **thalamus,** contain **ventral, medial,** and **lateral nuclei** and the **geniculate bodies.** These are synaptic relay centers for sensory impulses. The regulators for expression and emotion are located here. An electrode inserted into the thalamus area of a cat's brain will produce all of the spitting, clawing, hair-on-end manifestations of rage. It is believed that the lateral and ventral tracts for motor function, which are seen in the spinal cord and hindbrain, never extend forward into the diencephalon area.

The top of the diencephalon remains thin-walled, and the epithalamus bears only the remains of the dorsal pineal evagination. The hypophysis, or infundibulum, is the ventral evagination whose origin has been referred to elsewhere. It is extremely important because it produces the adreno-corticotropic hormones (ACTH), the gonadotropic hormone (GTH), and the thyrotropic hormones. These are very important in maintaining general body metabolism because they stimulate the production of hormones in the adrenal and thyroid glands and in the gonads. The nature of many of the diencephalon cells suggests that they also secrete hormonal (endocrine) substances.

Optic Chiasma. The optic stalks that connect with the optic vesicles, budded off earlier from the prosencephalon, enter the brain in the base of the diencephalon anterior to the hypophysis. The optic stalks have tracts of fibers, some of which cross from one stalk to another (the optic chiasma) in order to co-ordinate the image seen by one eye with that of the other. In lower vertebrates the eyes are at the sides of the head and each sees independently. One of the evolutionary trends in mammals has been the movement of the eyes forward to face front. Since this causes the visual field of each eye to overlap, the crossing-over of fibers enables the brain to make sense of the images and at the same time give depth perception. Visual centers have been developed in the cerebrum to accomplish this.

Telencephalon. The ancient telencephalon was primarily devoted to the sense of smell. The association centers for other functions were located farther back in the brain. This telencephalic structure is known as the **paleopallium** ("very ancient walls").

As the association centers moved forward in the brain in the course of evolution, an upper area devoted to sensory association neurons, called the **archipallium,** developed. Ventrally a motor association area, the corpus striatum, devoted to muscle tonus and movements, developed. Finally a small area on either side of the archipallium began to expand to form the neopallium. This forward area of the brain grew to form the large cerebral hemispheres containing billions of cell bodies of association neurons that enable the mammals to think and to solve problems, and also to direct or modify the reflexes that take place in the posterior part of the brain.

Embryonic development of the human brain reflects the evolutionary developments we have traced here. From the five-vesicle stage the two telencephalic vesicles begin to expand upward, forward, and backward over the brain stem at about the seventh week. The thickening area forms the corpus striatum in the base of each cerebral hemisphere. The dorsal wall in the center is quite thin, and drops down between the hemispheres, carrying choroidal tissue from the diencephalon inside with it. This is the remnant of the archipallium, the **hippocampus.**

Above the hippocampal area, nonolfactory sensory impulses enter the neopallium from the thalamic region. The cerebral cortex proliferates extensively, and finally a new set of fibers, the **corpus callosum,** arises and crosses between the two hemispheres to co-ordinate them. The only olfactory area is the pyriform cortex lying below the corpus striatum. The corpus striatum forms the caudate and lentiform nuclei.

It is said that in the period from seven weeks to full term the primary cerebral cortex develops into six recognizable layers of cells. The foldings of the cerebral cortex are quite regular in appearance, even among different races of humans. The functions controlled by the various areas have been mapped rather extensively.

The spaces remaining in the cerebral hemispheres are the first and second ventricles or telocoeles, which interconnect via the **foramen of Monro.** Medially they communicate with the third ventricle of the diencephalon.

The early flexing of the entire brain stem in human embryos makes relating the brain anatomy to that of other vertebrates difficult. Three flexures are recognized during development. The first is the apical flexure that bends the telencephalon around the notochord end. The next is in a reverse direction, dipping ventrally at the metencephalon region to give the pontine flexure. The last is a dorsad bend in the neck region, called the cervical flexure.

Cranial Nerves. Traditionally twelve cranial nerves have been listed for most vertebrates, and

TABLE III

The Cranial Nerves

Area	Name	Serves	Function
Diencephalon:			
0.	Terminal	skin	sensory
I.	Olfactory	nose	sensory
II.	Optic	eye	sensory
Mesencephalon:			
III.	Oculomotor	eye muscles	motor
IV.	Trochlear	eye muscles	motor
Myelencephalon:			
V.	Trigeminus		mixed, sensory, and motor
	a. Profundus (ophthalmic)	eye socket	
	b. Maxillary	upper jaw	
	c. Mandibular	lower jaw	
VI.	Abducens	eye muscles	motor
VII.	Facial	taste, face muscles	mixed
VIII.	Auditory	vestibule, cochlea of ear	sensory
IX.	Glossopharyngeal		mixed
	a. Superior	external ear	
	b. Inferior	tongue, pharyngeal muscles	
X.	Vagus (Parasympathetic)		mixed
	a. Superior	external ear	
	b. Inferior	pharynx, thoracic abdominal viscera	
XI.	Spinal Accessory	viscera, with X.	motor
XII.	Hypoglossal	tongue	motor

this is apparently the primitive number although some vertebrates show only ten.

The twelve nerves were assigned Roman numerals in the sequence in which they are located in humans. This did not take into consideration, however, that some vertebrates have a terminal nerve at the anterior end. To avoid confusion, this terminal nerve was given the number zero.

The primitive condition probably consisted of a dorsal nerve that carried afferent sensory fibers (those that went to the spinal cord) and possibly some visceral motor fibers such as the sympathetic fibers. The ventral roots were probably entirely separate somatic motor nerves. In the primitive vertebrates there was little deviation from the "straight tube" brain. Nerves would have come off segmentally in the anterior region to supply the body segments and the pharyngeal gill slits. The changes in body, with reduction or elimination of gill slits, plus the bending and bulging of the head region of more recent animals, obscures the origins of the cranial nerves.

Olfactory Nerves. The olfactory nerves are unusual in that they do not have cell bodies located in the brain or in ganglia, as most of the cranial nerves do. Instead, the cell bodies lie in the sensory epithelium of the nose, and their axons grow back through the skull to synapse with association neurons on the olfactory bulbs or olfactory area of the brain.

Optic Stalk. The structures identified as the optic nerves are actually the optic stalks of the brain, which grew from the telencephalon as the optic vesicles moved out to become the eyes. When the eyes form, the round vesicles invaginate so that the outer half of the vesicle is pushed against the inner half, like a cup within a cup. The innermost part forms the pigmented layer; the outer part becomes the **retina,** the sensory layer. As the vesicles fold in, a crease or groove is left along the bottom of each cup, and the stalk eventually grows around the crease, leaving a lumen (tube) inside the stalk.

Nerve cells develop in the retinal layer, but instead of the axons growing out the back of the layer they grow forward. The axons run across the retina, enter the lumen of the eye stalk, and grow back toward the brain proper, to form the optic nerve. Where the axons enter the eye stalk there are no sensory cells, resulting in a blind spot.

Blood vessels also grow through the stalk; from the brain area they extend out into the eyeball, but they atrophy before birth. Neurons in the nasal epithelium and in the retina would be comparable to the sensory cell bodies in the dorsal root ganglia.

As in the optic cup, the eye lens forms from the invagination of a vesicle of ectoderm in front of the optic cup. When the lens vesicle pinches off, mesenchyme cells from the crest are in the process of covering the brain and eyeballs with a vascular layer (**choroid coat**) and then a tough, fibrous layer (the **sclerotic coat**). The cornea is a clear region of epithelium left from the formation of the lens vesicle, and is continuous with the sclera. The iris and ciliary muscles are formed along the edges of the optic cup by invading mesenchyme. **Aqueous humor** is secreted into the space between the lens and cornea, while the **vitreous humor,** a thick jelly, is formed between the lens and retina. The surface ectoderm forms folds that grow down to become the eyelids. They fuse and later reopen at the point of previous fusion at about seven months.

Vagus Nerve. The vagus nerve arises as the tenth cranial nerve in the posterior medullar region. Joined by fibers from neighboring cranial nerves, the vagus leaves the skull and grows down to form terminal ganglia in the region of the viscera. The vagus nerve forms the major portion of the parasympathetic system, which is part of the autonomic nervous system.

Autonomic System. We have already discussed the segmental spinal ganglia along the spinal cord in the chest and abdominal region which form the sympathetic nervous system. Fibers from these ganglia innervate the visceral organs, and the hormone adrenin or epinephrine provides stimulation at the neuro-muscular junction (the point at which neuron meets effector cell).

The parasympathetic system is primarily cranial in origin and has terminal ganglia on or near the viscera. The neuro-muscular junctions in this system are stimulated by acetylcholine, the usual synaptic fluid. Strangely, the sacral spinal nerves (the "tail" region) are activated in this manner also, so that the sacral spinal ganglia are part of the parasympathetic system.

It can be seen, thus, that the viscera are controlled in mammals by two systems. Function is maintained by a balance between the two, since their actions are antagonistic. The sympathetic system speeds up the heart rate and raises blood pressure. It causes raising of the body hair and flow of the sweat glands. It also suppresses the activity of the digestive organs and relaxes the bladder. The parasympathetic system produces opposite effects. It slows the heart, lowers blood pressure, and decreases sweat. The activity of the digestive tract increases, and the bladder and lungs contract.

The human nervous system demonstrates its heritage from simpler animals by the basic brain stem and the spinal cord and by its simple reflex arcs. The most fundamental functions that require no thought are routed through the hindbrain area. As the complexity of the task to be done increases, the nervous system routes stimuli through centers in the more anterior portions of the brain. Where tasks require perception and evaluation of a situation, and call for a variety of responses, the entire brain and nervous system is called into action. This separation of levels of action is good economy, for it sorts the incoming stimuli and outgoing motor responses so that the cerebrum need deal only with complex associations. The cerebral area was the last to arise in evolution, and is the last to develop in the embryo.

THE DIGESTIVE SYSTEM

Although the gut is not conspicuous in early embryonic stages its development influences the development of the circulatory system, as we shall see shortly. You recall that the vertebrate gut originates from the endoderm of the primitive streak. The embryo then becomes raised and the somatopleure folds under it at the head and then at the tail. The foregut and hindgut become closed tubes, leaving the midgut region open into the yolk sac. The allantois develops as a diverticulum of the hindgut, and the open gut region is progressively reduced until only the body stalk, or umbilicus, remains open. The gut then begins to differentiate by forming diverticula at various levels.

In order to understand the significance of the developing human digestive structures, it is helpful to examine the anatomy of the gut of lower vertebrates. The primitive prechordates probably resembled *Amphioxus* to some extent. These animals have a subterminal mouth without bony jaws. The pharyngeal region is expanded into a wide cavity perforated by slits that allow water to enter through the mouth and exit through the slits. This is a feeding mechanism, and in order to trap small particles the tissue in the area secretes mucus. In the floor of the pharynx is a ridge of such tissue, called the **endostyle,** which helps to twist the food into a rope of mucus which will be carried through the rest of the digestive tract. The slits, called **visceral clefts,** are supported by visceral bars of cartilage embedded in the tissue between them. The bars usually have paired ventral, lateral, and dorsolateral segments.

Visceral Arch Derivatives. The ancient fishes were agnathan (without jaws). Gradually the fish began to eat larger pieces of food than were brought in by pumping water, and this was probably done by bracing the mouth against the first visceral bar. Slowly the first arch cartilages migrated forward until the upper ends of the cartilage moved dorsad to form the upper jaw, and the lower portion became the lower jaw. Certainly this must have been a great advantage in competition for food.

In humans, a similar sort of formation takes place between the fourth and eighth weeks. The upper parts of the mandibular arch cartilage begin to form in the cheekbone regions and grow toward the center of the face. The lower parts of the cartilage grow toward the center to form the lower jaw. These later become replaced by bone. Failure to fuse in the center produces such defects as cleft palate and cleft ("hare") lip.

In progressing from primitive fish to modern vertebrates, many modifications of the other visceral arches occurred. Part of the primitive first arch has been modified and has migrated so that an upper portion becomes the incus (anvil) bone and a part becomes the malleus (hammer) bone of the middle ear. The uppermost part of the second (hyoid) arch joins these as the stapes (stirrup). These three bones help to transmit sound from the outer ear shell to the inner ear sensory area.

The basic vertebrate body plan includes at least seven visceral arches with pouches between them and slits opening to the outside. Some ancient animals had more, but remains of seven can be traced in all vertebrates and can be found during embryonic development in mammals.

The first arch and part of the second, as we have seen, became included in the skull. The remaining arches supported the gills in fishes but became modified as the number of gill slits were reduced during evolution. The lower portion of the second arch, the hyoid, moved forward to provide support for the tongue.

When amphibians began to develop lungs for air-breathing, visceral arches were modified to form the larynx, the support for the tracheal opening of the lungs. At first, the seventh arch contributed to the tracheal rings and arytenoid cartilage, which supports the tracheal opening (glottis). This became further strengthened by the cricoid cartilage.

When vocal cords began to appear in mammals, the sixth arch formed the epiglottis cartilage to close the trachea from above. Finally the third, fourth, and fifth arches contributed to the larynx. The third disappeared in higher mammals, leaving the thyroid cartilages (the word means "shield shaped" and does not refer to the thyroid gland) fused, to anchor the tongue and support the epiglottis.

Pouch Derivatives. The pouch tissue preceding each visceral arch has also been modified. Authorities differ in numbering the visceral arches and their pouches, but it is worthwhile to examine them in an established sequence. The mouth, which probably was terminal in the ancestral animals, moved back as the first arch moved forward to form the jaw. Therefore comparative anatomists assign Pouch I to the mouth region. Pouch II moved upward to form the middle ear cavity and the **eustachian tube,** which connects the middle ear to the throat. The third through the seventh slits and their pouches virtually disappeared, but are seen in embryonic development.

In human embryos, four pouches appear early in development. The slits do not normally open, but are closed only by thin layers of ectoderm and endoderm back to back without intervening mesoderm. Occasionally the third or fourth slits do open, causing the defect known as cervical fistulae, holes leading to the throat or mouth. Embryologists do not number the pouches from the evolutionary standpoint, counting only actual existing structures.

Thyroid Gland. The sides of the first pouch seen in human embryos form the middle ear and lining of the eardrum. The external tissue between the mandibular and hyoid arches becomes the external ear shell. The floor of that pouch contributes to the tongue, and the thyroid gland begins as an evagination in the center of the developing tongue on the floor. Eventually, the thyroid lies beneath the fourth pouch region, losing its connection with the tongue as the interior pharynx and face grow forward. By the fourth month of development, the thyroid tissue has elaborated and fused with tissue from the area behind the fourth pouch and begins to secrete thyroxin, a metabolic hormone. Iodine is concentrated by the thyroid; iodine deficiency in the diet will cause the gland to enlarge, forming a fibrous growth, a goiter. In the midwestern United States, where no ancient seas had ever deposited iodine in soil and little seafood was consumed, goiter was common in past years. Now iodine is added to commercial table salt, and goiter is rare in those regions.

Tonsils. The second pouch in the human embryo provides the entrance of the pharynx with a ring of lymphoid tissue, the tonsils. The pharyngeal tonsils (adenoids) lie above and behind the roof of the mouth (the hard palate). The sides of the pouch become the palatine tonsils, the ones visible in sore throats and often removed by surgeons. A lingual tonsil far back on the base of the tongue completes the circle. In other vertebrates tonsil tissue may form from other pouch areas.

For years it was assumed that the tonsils were unnecessary vestigial remains from the ancestors. Recent research has indicated that they are important lymphoid tissue, contributing to the bodily defense against foreign proteins such as bacteria.

Carotid Bodies. Another pair of organs that originate in the ventrolateral region of the second human pouch are the carotid bodies. These bodies originate in the aortic arches that supplied the visceral clefts and the arch tissues, rather than from gut diverticula. They receive nerve cells from the autonomic ganglia of the vagus nerve, and meter the blood pressure and carbon dioxide content of the blood. Stimuli are relayed to the medullary centers for control of blood pressure and rate of respiration.

Thymus. The thymus gland begins as a pronounced ventral sac at the level of the third pouch and by seven weeks in the human it has detached from the pharynx. It becomes bilobed and the lower ends attach to the lining of the heart cavity, the **pericardium.** This pulls the thymus down as the pharynx grows, and it comes to lie posterior to the thyroid gland.

The thymus is extremely important as the primary source of the lymphocytes in the bloodstream. The primordial lymphoblasts are apparently formed in the embryonic thymus after about eight weeks and travel through the developing bloodstream and the lymph channels. Some settle in the spleen and bone marrow while others form the lymph nodes, such as tonsils, spaced irregularly through the body in the mesodermal lymph ducts. The lymphoblasts give rise to both fixed and free microphage cells and lymphocytes that attack foreign bodies in the lymph or blood.

The embryo has no antibody reaction for much of the gestation period. Some foreign substances will cause a reaction at about the seventh month, but others do not react until after birth. It is said that at a certain point in development the fetal thymus "registers" all of the proteins present, and anything introduced after that will be considered foreign. Transplanting experiments have verified this to some extent. Errors in the system lead to medical problems where allergies to the patient's own tissue are found. Some antibodies from the mother can pass through the placenta, which complicates determining what is being formed by the fetus. The thymus is large in children, but virtually disappears in adults.

Parathyroid Glands. The dorsal portions of the third and fourth pouches in humans give rise to the parathyroid glands. They separate from the pharynx at about the same time the thymus does and come to lie above and below the thyroid. The parathyroids play a critical role in balancing calcium and phosphorus ions in the body. Removal of the parathyroids causes death, and hence they must be carefully separated from the thyroid if the latter must be removed because of malignancy.

Postbranchial Bodies. Small nodes of tissue may arise from other areas of the pouches. In mammals they migrate to the thyroid area, but may lie more posteriorly in other vertebrates. The function of the postbranchial bodies is unknown. The word "branchial" refers to respiration; since the pharyngeal arches bear gills in fish, they are also called branchial arches.

Lungs. Primitively, the fish often developed some type of air sac or swim bladder that formed from a diverticulum near the most posterior pharyngeal pouch area, perhaps from the merger of several former gill slits. Air sacs may have developed as an aid to buoyancy when fish invaded fresh water, and in some fish the bladders aid in hearing vibrations. The attachment to the gut may have originally been ventral, but is either dorsal or lost altogether in modern fish. The lungs may or may not be homologous (of like origin) with the swim bladder. They do not have comparable blood supplies; this suggests that they are not homologous.

The lungs originate as a single ventral diverticulum of the gut. The lung bud immediately begins to divide, forming two main branches, the **bronchi.** In amphibians, the first animals to use lungs for respiration, the lungs are very simple sacs. Their moist skin provides most of the needed surface for oxygen exchange. Branching of the bronchi into bronchioles, and subdivision of the sacs, gradually increased the surface necessary for complete respiration in the lungs of reptiles, birds, and mammals.

In humans, the right bronchus has three main lobes; the left has only two, probably because the heart is flexed to the left. During the first four months the main bronchial divisions occur. From the fourth to the sixth month the bronchi branch repeatedly to form the tiny tubules, and the epithelial cells develop cilia. From the sixth month on, tiny air sacs, the **alveoli,** form at the tips of the tubules. Blood capillaries invade the alveolar area and the fetus is sufficiently supplied to survive birth during the seventh month. Growth and proliferation continue even after birth. Detergent-like secretions called **surfactants** are produced as birth approaches. These cause the lung epithelium to be freed of mucus and amnionic fluid. The surface cells also must be kept moist to prevent hardening of the cells. Respiratory failure is a common cause of death in premature infants.

Esophagus. The gut originates as an undifferentiated straight tube, but as growth takes place the tract lengthens greatly. Differentiation occurs at various locations. At first the esophagus is short, connecting the pharynx with the stomach, but it soon lengthens. It remains relatively simple in structure compared to the specialization of other gut regions.

Stomach. In the upper abdominal region the stomach begins to grow so that it bulges to the left side and turns. The mesenteries that support the entire gut tube dorsally and ventrally at first become twisted because of the growth and twisting of the gut. The ventral mesentery breaks down completely below the stomach. A large pouch called the **omental bursa** forms from the dorsal mesentery. Later, mesenchyme cells settle in the mesentery to form the spleen, a center of lymphocyte production.

Intestine. The narrow gut region between the stomach and umbilical stalk is the small intestine, which also begins to twist. In order to accommodate the rapidly increasing length, it bulges out into the extra-embryonic coelom beside the yolk sac during the first ten weeks. Below the body stalk the intestine widens abruptly into the large intestine or **colon.** A pouch called the caecum occurs at the juncture; it bears a smaller diverticulum, the appendix, which has no known function. The caudal end of the intestine bears the bladder and then forms the rectum leading to the anal opening. Later, the bladder separates from the gut and forms its own duct and opening.

Digestive Glands. At the juncture of the stomach and small intestine (the duodenal region) several diverticula arise early in development. First the hepatic bud forms and divides into two main branches, one becoming the lining of the gall bladder and bile duct and the other the liver. The liver bud then divides into two lobes and becomes heavily invested with mesenchymal cells, forming a prominent bulge in the embryo. The liver extends up to the septum transversum, a muscle layer that grows from the ventral body wall to help form the diaphragm partition between the thoracic (chest) and abdominal cavities.

At the level of the hepatic bud two other buds, the **pancreatic primordia,** appear. The dorsal pancreas is the larger, growing up along the stomach. The ventral pancreatic duct moves up and becomes a branch of the hepatic duct as it grows out. The rotation of the duodenum and stomach causes the ventral pancreas to come to lie beneath the dorsal pancreas, instead of on the opposite side of the gut. The tissues fuse and in humans the duct of the dorsal pancreas joins that of the ventral pancreas so that the pancreas empties via the ventral duct into the common bile (cystic) duct. Isolated groups of cells separate from the pancreatic tissue internally to form the **islets of Langerhans,** secretory cells that produce insulin for the metabolism of sugar.

Cellular differences characterize the development of the various digestive organs. The esophagus is lined with stratified squamous epithelium. In the stomach, gastric glands form and several types of secretory cells occur there and in the intestine. The villi, which give added absorptive surface, form in the intestine early in embryonic development.

Some of the digestive enzymes are present in the gut by the end of the fourth month; the epithelia have begun secretion of proteolytic enzymes and trypsin. At birth the intestine is filled with a greenish mucous substance, meconium, which is composed of secretions from the liver and intestinal glands, and swallowed amnionic fluid.

Failure of any portion of the gut to differentiate properly will affect the function of the entire tract. Occasionally the umbilical opening into the gut fails to close, causing feces to exit through an umbilical fistula.

THE CIRCULATORY SYSTEM

Knowledge of the primitive pattern of the vertebrate visceral arches makes the patterns of the developing circulatory system more understandable. The vertebrate heart is ventrally located and pumps blood forward into the neck and head region by way of a median ventral aorta. In primitive vertebrates and in modern fish paired vessels are given off which lead to the gill arches on each side. These are the **aortic arches,** which pass up along each side of the gill slits and then enter paired dorsal aortae which lead caudally. The dorsal aortae merge into a single median dorsal aorta back of the gill region. The dorsal aorta continues down to the tail region, giving off paired vessels to the visceral organs.

Aortic Arches. In gilled animals the aortic arches develop loops that encircle the gill slits, small arteries that cross between loops, and capillaries that extend into the gill filaments for oxygen exchange. The blood sent into the gills by the heart pump would thus be deficient in oxygen. As the number of gill slits was reduced and the arch cartilages modified, the circulatory patterns also changed. Six aortic arches are basic to all vertebrates; the first arch becomes modified to supply the jaw and the second supplies the middle ear and tongue. The third arch contributes to the carotid arteries to the head. The fourth arch becomes the main arch passing blood up and back in the dorsal aorta. The fifth arch tends to disappear, and the sixth becomes modified to supply the lung.

The advent of the lung behind the last gill arch in certain fish enabled the vertebrates to move onto the land and begin the amphibian line of animals. A side branch from the sixth aortic arch was diverted to supply the lung. The sixth arch plays an important part in embryonic circulation and in changing to lung respiration at birth, as we shall see.

Development of the human circulatory system begins very early in the embryo; the bulge of the growing heart can be seen when the brain is still in the three-vesicle stage. It will be recalled that clusters of mesodermal cells gather into blood islands in the extra-embryonic area at the primitive streak stage. Cells in the clusters gradually merge, organizing themselves into tubules of endothelium that will form the linings of the future blood vessels. Other cells enclosed by the tubules form the future circulating blood cells.

The tubules on either side of the embryo proper form a network leading into two main tubes, the **vitelline veins.** These veins lead toward the midventral area and meet just in front of the intestinal portal where the headfold is causing the gut to become closed. The paired vessels grow forward side by side, and gradually the adjacent lateral walls merge to form a single tube, the heart.

Anterior to the pericardial (around the heart) area the tubes diverge again, forming two ventral aortae which pass up the first aortic arch in the mandibular area. These become the so-called **ascending aortae.** They then loop back as paired dorsal aortae, running along either side of the developing neural tube. They gradually fuse into a single aorta, from the caudal region forward. The large ventral umbilical arteries and paired dorsal intersegmental arteries are given off by the dorsal aorta.

As the embryo grows forward, the second arch is formed at the aortic sac region, immediately in front of the developing heart. The formation of a neck in the higher vertebrates causes the heart to be displaced caudally. The third and fourth arches appear as this displacement occurs; simultaneously the first and second arches disappear, leaving their connections represented only by the maxillary, stapedial, and carotid arteries. The fifth arch appears only briefly and the sixth appears as the lung bud begins to develop. The sixth arch immediately gives off descending branches to the lung region. After this stage, segments of the basic arches begin to disap-

pear, progressing to the adult pattern. It is interesting to note that in the mammalian line of evolution the left fourth aortic arch becomes the main branch leading to the dorsal aorta, but in birds the right fourth aortic arch becomes the main branch.

The venous channels arise at the same time the arteries are organized; the two systems can only be distinguished at first by direction of future blood flow. Later the arteries are invested with muscular layers because they must sustain the heavier pressure from the pumping heart. The veins are thinner-walled and develop valves on the inside. Since the blood passes through capillaries into veins, the pressure in the veins is relatively low. The valves prevent backflow of the blood returning to the heart. The heart valves are undoubtedly of venous tissue origin.

The Venous System. The vitelline veins, which approach the heart laterally, form the central part of an H-shaped system. The upper arms become the anterior cardinal veins, and the lower arms the posterior cardinal veins. In *Amphioxus* this remains the fundamental system. In all the vertebrates a subintestinal vein, ventral to the abdominal viscera, begins to take on importance.

The developing liver, just below the heart, forms a network of many blood channels (sinusoids). These merge with the vitelline veins and then connect with the intestinal supply so that a single median venous channel is formed. The **hepatic portal vein** leads to the liver and the **hepatic vein** leads from the liver to the heart. In lower vertebrates another portal system, the renal portal vein, leads to the kidneys. The blood is passed through the kidney capillaries and then renal veins carry it into the ventral abdominal channel. This system disappears above the amphibian level.

The anterior cardinal veins form new connections above the heart so that a single vessel, the **anterior vena cava,** leads into the heart. Posteriorly, the hepatic vein eventually forms a new path and becomes the **posterior vena cava,** joining the anterior vena cava at the entrance of the heart.

Changes in blood vessel pathways undoubtedly result from evolutionary changes in blood pressures at various regions. In gill breathers, venous blood is pumped under low pressure into the gills for aeration in the capillary beds because high pressure would rupture the capillaries. Excretion also occurs in the gills to some extent. With the shift to lung breathing, venous blood is pumped from the heart into the lungs and returns to the heart to be pumped under high pressure through the body. Pressure is reduced by the time the blood reaches the kidney capillaries. Many of the transitions from the primitive pattern to the modern can be observed during embryonic development. The venous system is harder to trace, however, because it tends to break up into sinusoids and to develop abnormal connections.

The Heart. The primitive prevertebrate animals did not have a true heart; a simple collecting sac, the sinus venosus, led directly to the ventral aorta, which pumped by muscular contractions.

The basic vertebrate heart consists of four specialized regions of the pumping tube; from posterior to anterior they are the sinus venosus, the atrium, the ventricle, and the conus arteriosus (or bulbus cordis as it is sometimes called in human embryology).

It has often been said that fishes have a two-chambered heart, amphibians a three-chambered heart, and mammals a four-chambered heart. This is a misleading statement, however, for such a system of counting is based on the presence or absence of partitions in the atrium and ventricle alone.

The sinus venosus and conus arteriosus are extremely important in lower vertebrates and in the evolution of the heart. Some authorities include a fifth region, the truncus arteriosus, but this is apparently histologically related to the aortae rather than the heart.

The primitive fishes had the four basic chambers lying in a straight horizontal line, collecting the blood in the sinus venosus from common cardinal veins on either side. The blood passed forward from the sinus into a thin-walled atrium, then into the thicker-walled ventricle, and out the conus into the ventral aorta or truncus. The blood pressure was low, so that the blood could pass through the gill capillary beds for oxygenation without rupturing them.

Folding. Modern fishes began to develop folding in the heart tube, with the sinus venosus and atrium being pushed forward so that they were dorsal to the ventricle and conus. This increased blood pressure by gravity flow; at the same time some of the basic pharyngeal arch blood pathways were abandoned and their circulation directed elsewhere. As the kidney assumed more of the excretory function, an increase in blood pressure helped to provide the necessary circulation in the posterior region.

Lung Circulation. The advent of the lung necessitated great changes in circulatory systems, and blood was diverted from the sixth arch to supply the lungs. A greater problem, however, was the mixing of the venous blood, carrying carbon dioxide, with the oxygen-carrying blood returning from the lung. The heart previously had handled only venous blood when the gills provided oxygenation.

Partitions were developed from the fibers and cushions of the chamber walls to help keep the venous and oxygenated blood separated. Division of the atrium, with the sinus leading to the right side, allowed the pulmonary veins to enter the left side. The ventricle remains undivided in the amphibians, although fibrous ridges help to guide the two types of blood. The conus developed a split, forming a right-hand trunk which led directly to the lungs, and a left-hand trunk which led to the systemic or fourth arch. Further folding pushed the atrial area forward until it lay anterior to the ventricle region in the adult.

The next principal change in the heart consisted of a drastic reduction in the sinus venosus, coupled with the direct entry of the pre- and post-caval veins into the right atrium. The sinus venosus contained the nervous center that controlled the contraction of the entire heart. Since this center is required even though the sinus cavity is not, the center became located on the atrium. It is known as the **sino-atrial node** or the **pacemaker.**

Reptiles clearly show the various transitions leading to the bird and mammalian systems. Some reptiles have an incomplete partition in the ventricle, although in the crocodiles it is complete. This directs pulmonary blood through the left atrium and left ventricle and out the left aortic arch. Fibers from the sino-atrial node extend down into the partition to form the atrio-ventricular node.

The conus region is split into three aortae: left systemic arch, right systemic arch, and the pulmonary artery, in reptiles. The right atrium receives blood from the vena cava. Venous blood passes on through the right ventricle and out either the pulmonary artery or the right systemic arch, which then curves to the left side. Since it is not desirable for the left side of the body to receive blood deficient in oxygen and high in carbon dioxide, an opening, or foramen, allows some oxygenated blood to flow from the left ventricle to the right, at the base of the aortae, to supply some oxygenated blood.

The mammalian heart modified the reptilian pattern by dropping the inefficient right aorta and depending on the left to carry the entire load. It, in turn, divides but the right branch supplies only the right side of the head and right arm, while the left branch carries the burden of supplying the body as well as the left side of the head and left arm.

Warm-bloodedness. Mammals and birds require a more efficient heart because they maintain a constant temperature (homeothermic). Other vertebrates are said to be cold-blooded animals because their body temperatures are dependent on the temperature of their surroundings (poikilothermic). Cold-blooded animals metabolize according to the temperature so they feed and move around only when it is warm. Their oxygen needs are very low when the temperature is low. It has recently been discovered that some of the cold-blooded invertebrate insects, and some reptiles, are able to modify their body temperature to some extent. Some insects fan their wings before flight, which raises their body temperature and speeds up metabolism. After flight and feeding, the insects rest; food is conserved by lowering the metabolic rate and body temperature.

Birds must feed almost constantly to meet the demands of flight and a higher body temperature. Their body temperature ranges from about 102° F. to 105° F.; a parakeet perching on your finger will have noticeably warm feet. The heart rate is nearly 1000 beats per minute, and the air sacs have been extended into the hollow bones to supply oxygen for the high metabolic rate. The heart rate of humans averages about 80 beats per minute and the normal body temperature is 98.6° F.

Primates add the stress of vertical posture to the demands of warm-bloodedness. The heart must begin functioning early and continue efficiently under any situation. Developmental defects that impair this efficiency severely restrict physical activity.

Embryonic Circulation. The human heart must function from almost the beginning of organogenesis in order to supply the high embryonic metabolic rate and to circulate blood through the placenta for nourishment and excretion. It is believed that circulation begins as early as the fourth week, when the embryo is only about 4 millimeters long. (The placenta is quite a bit larger than the embryo until about the fifth month.)

Although the heart arises as a straight tubule of endothelium, it soon becomes invested with a layer of mesoderm, the epimyocardium. The two layers are separated by a reticulate jelly substance which is eventually invaded by developing muscle fibers from the epimyocardium. The mesodermal folds form the pericardial cavity and the dorsal mesentery which at first suspends the heart.

The same four basic regions that can be seen in the tubular fish heart are present in the human heart tube. The heart grows faster than the supporting region, and so the tube begins to bend and bulge to the left side. Finally it twists into a loop, which brings the sinus and atrial regions anterior to the ventricle. The conus (bulbus) region then leads from the posterior end of the ventricle, across the ventricle and atrium surfaces, into the truncus and the aortic arch. The pressure of the conus may help to form the partitions in atrium and ventricle by depressing the center areas.

Septa, which will form partitions to divide the atrium and ventricle into two chambers each, begin to develop, but remain incomplete until birth. An opening is found between right and left atria (the

foramen ovale), and another remains between the right and left ventricles (the interventricular foramen) during the fetal period.

The foramena provide for a circulation pathway adapted to oxygenating the blood in the placenta. The fetal lungs are merely developing buds that cannot handle a blood supply proportional to the embryo's needs after birth.

The blood in the fetus enters the right atrium via the superior and inferior vena cava, since the sinus becomes incorporated in the atrial wall. The blood may then pass down past the tricuspid valve into the right ventricle, but most of it passes through the foramen ovale into the left atrium. The partition actually is composed of two septa, each with a hole in it, side by side. The holes are not opposite each other, and when the atrium contracts, the blood passes through the first hole and pushes the second septum aside sufficiently to enable it to pass on through the second hole.

The blood that reaches the right ventricle is pumped out the pulmonary trunk to the lungs. When the lungs are small, the foramen between the right and left ventricle is open, to further reduce the blood passing out the pulmonary trunk. (The truncus contricts and finally forms two separate tubes, the pulmonary trunk and the aorta.)

Blood returns from the lungs to the left atrium and, with the blood that comes through the foramen ovale, passes through the mitral valve into the left ventricle. From here it is pumped out the aortic trunk to the body.

From the standpoint of comparative anatomy origins the main aortic arch is the fourth arch. The pulmonary trunk is the sixth arch plus a side branch. During development, the ductus arteriosus, the end of the sixth arch, connects the pulmonary trunk to the descending aorta, again reducing the amount of blood sent to the lungs. Only enough blood is sent to the lungs to establish the pathway and nourish them.

Changes at Birth. At birth, oxygenation is suddenly stopped in the placenta. At the same time, the ductus arteriosus constricts, shunting all the pulmonary trunk blood through the lungs. This increased flow of blood into the left atrium causes pressure against the interatrial septa, preventing blood from the right atrium from passing through the two holes of the foramen ovale. Since the interventricular septum was completed much earlier, blood is now forced into the normal adult pathways.

Failure of the ductus arteriosus to constrict at birth causes the infant to circulate unoxygenated blood through the body, and the reduced volume passing through the lung may in turn cause the foramen ovale to remain open. These two occurrences can cause the so-called "blue baby" condition in which the blood does not receive sufficient oxygen to form the red oxyhemoglobin and supply the metabolic needs of the body.

Valves also develop in the bases of the pulmonary and aortic trunks. These are the semilunar valves, which have cusps (cups) in them that prevent blood from flowing back into the heart after completion of the contraction that pumped it from the ventricles. The valves between the atria and ventricles are similarly placed so that they close and prevent backflow. Damage to any of these valves can seriously reduce the efficiency of the heart. German measles contracted by the mother during pregnancy can cause such damage to the infant.

The Lymphatic System. The lymphatic system consists of a group of channels that develop at the same time the circulatory system is forming. The channels are quite regular in most amphibians, but much less so in mammals, forming as mesenchymous spaces that become confluent.

The lymphatics serve to return much of the fluid that leaks from the arterial capillaries because of high blood pressure or osmosis, and is not collected by the venous capillaries. The lymph channels also carry large fat molecules that do not seem to be able to pass into the bloodstream in the intestine.

In mammals, lymphatic channels, the **iliacs,** in the legs, drain into the main abdominal pool or vessel, called the cysterna chylii. Other channels from the abdomen join there, and the main channel travels through the chest as the thoracic duct. This receives channels from the neck and then turns, at about the left collarbone, and empties into the left subclavian vein just before it enters the superior vena cava. In this way fluid is returned to the bloodstream. In some mammals, although not in humans, the thoracic ducts are paired.

The lymph nodes are not a functional part of the lymph channels but are formed when lymphoblasts from the thymus gland lodge in the various sections of the channels and begin their protective functions.

Somites. Caudal to the developing brain and heart, blocks of tissue called somites form along both sides of the neural tube. Somites are added a pair at a time at the caudal end during early formation and growth. The number of somites is actually a more accurate determination of the age of an embryo than the supposed number of hours of incubation.

Generally accepted standards for the chick give a 24-hour embryo four pairs of somites; a 26- 27-hour chick has seven pairs of somites; at 30 hours there are ten pairs of somites; at 36 hours there are fourteen pairs; on up to thirty-five somite pairs which occur at about 85 hours. After that, the body develops detail which tends to obscure the somites.

Mammals are usually measured in millimeters of length from crown to rump. This dimension is used because of the C-shaped position assumed by the body very early in development. Mammals add somites caudally in the same manner seen in chicks. At about twenty days a human embryo has ten somite pairs; at about four weeks there are twenty-five pairs and the crown-rump length is 3.6 to 4.5 millimeters.

After the somite blocks form, the tissue differentiates according to the tissue adjacent to it. The dorsolateral portion, nearest the body surface, becomes the **dermatome** which underlies the epidermis and forms the blood system and muscle layer of the skin. The portion next to the neural tube is the **sclerotome** which forms the vertebrae, skeleton, and skeletal muscle. The most ventral portion of the somite, the **myotome,** forms the coelomic cavity lining, the muscles of the gut, and tissues of the circulatory system and part of the gonads. The ventrolateral portion forms the nephrotomal plate, which forms the kidneys and reproductive ducts. Since segmentally arranged tissues sometimes are given the suffix -*mere,* the upper part of a somite is the epimere, the middle part is the mesomere, and the ventral part is the hypomere. Either set of terminology is correct, and both are sometimes applied together.

This concludes our consideration of development and of morphology. It is impossible to study one without reference to the other, and to do otherwise is to deny the student a more complete understanding of the dynamic aspects of anatomy.

CHAPTER 8

THE ANIMAL KINGDOM

A fascinating variety of living organisms exist which show an incredible range of size, shape, and complexity. These organisms also show wide differences in habitat, environment in which they live, and their methods of coping with that environment.

In surveying the animal kingdom, we inevitably encounter the names that have been given to the various groups by scientists over the period since Linnaeus developed the system of nomenclature presently in use. Latin was the mainstay of higher education during this time and the names given were in Latin, although they sometimes had Greek roots. These names are frequently descriptive of the animal. Since the study of Latin and Greek is not as widespread now, however, the names often seem unintelligible or ridiculous to us. An unabridged dictionary gives the derivation of many of these words, offering insight into the reason a particular name was given to a particular animal.

Animals are divided into a few major groups on the basis of several factors. Appearance is the first factor that was used, but we have come to know that appearances may be deceiving. Therefore, we must examine the various animals more closely to determine the actual similarities in pattern of development, structural complexity of the adult, and the way they carry on bodily functions.

The separation into groups is based upon the presence or absence of certain characteristics. The first division of the Kingdom Animalia asks the question whether the animal is single-celled (or acellular; without cellular divisions), or whether it is multicellular. This gives us two subkingdoms, the Protozoa and the Metazoa respectively.

Symmetry. Another criterion is the type of body symmetry of the larval and adult forms. The most primitive body shape is presumably spherical; all axes through the center are alike. **Radial symmetry** is somewhat similar, with like halves along any two diameters at right angles to each other. There may be a difference between the oral and aboral ("the other end") surfaces. **Biradial symmetry** is a further modification, in which halves in the sagittal plane differ from halves in the transverse plane. A sagittal plane cut in a fish, for example, is lengthwise, passing from the dorsal (back) surface to the ventral

(abdominal) surface. A transverse plane cuts the fish in half crosswise from dorsal to ventral surface. A fish is **bilateral** in symmetry. The more advanced animals usually have bilateral symmetry, wherein the body can be divided into identical halves only by a sagittal plane. They show dorsal-ventral differences, and anterior (front)-posterior (rear) differences.

A problem that develops when scientists try to evaluate the primary characteristics of an animal when it is first discovered is that a larval type may be mistaken for an adult form, or that the larva of a given adult may not have been recognized.

Animals that are sessile or slow-moving frequently revert to radial symmetry after a motile bilateral or biradial larval stage has passed, and yet the level of complexity of the organs in the animal is considerably above that of the simple, primarily radial groups. Examination of the larva may show that it is bilaterally symmetrical, in which case the animal is said to be a member of the bilateral group that has become secondarily radiate.

Organs Present. The presence or absence of various organs or organ systems is also a key to the complexity of an animal group. Digestive and reproductive organs are the first to appear in the sequence of advancing complexity. In some radiate groups, and in the more primitive bilateral groups, the digestive organ is just a sac with only one opening, which must serve as both extrance and exit. A one-way gut, with separate exit (anus) is an example of a significant improvement in an organ system.

Tissue Complexity. Another criterion that is sometimes used in judging relationships is the complexity of the metazoan tissues. In discussing the development of the germ layers, we mentioned that some of the lower invertebrates did not develop mesoderm as a separate germ layer but formed mesenchymal mesoderm cells from proliferation by the ectoderm and sometimes by the endoderm.

Some authorities say that animals having only two actual layers of tissue, with mesenchymal mesoderm cells not truly organized as a tissue, are **diploblastic.** Those with a tissue grade of mesoderm are **triploblastic.** This concept has been misleading in many instances because "blastic" cells are usually considered to be developmental, not adult, forms. (You will recall this terminology from the chapter on embryology.) "Diploblastic" would then refer to those animals that show only two germ layers, ectoderm and endoderm, and "triploblastic" would refer to those animals that form a mesodermal layer at the same time the ectoderm and endoderm are being determined from the epiblast and hypoblast. The terms have wrongly been applied to the adults, which leads to a different interpretation. The group of animals that are embryologically diploblastic may not be the same group that are supposedly diploblastic as adults. Therefore authorities have recently tended to avoid using this criterion.

Body Cavity. Related to the primary germ layers is another criterion used in grouping animals, that of the presence or absence of a coelom, or body cavity. The more primitive metazoans generally are small-bodied and move about rather slowly, if at all. The body basically consists of an inner layer and outer layer, with mesenchyme forming the bulk that lies between these layers.

In the simplest animal forms, the middle layer may be principally gelatinous with wandering mesenchyme cells, or it may be organized into muscle strands or bands in some animals. Because most of these animals are small and slow-moving, bulk is not a problem. If larger animals were to be constructed in such a solid manner, of course, several problems would be presented. The weight of this mass would be considerable, making movement very difficult—even impossible without enormous muscles. Distribution of food and wastes would also be a great problem, for there is no organized circulatory system in the simpler animals.

Some of the lower invertebrates begin to show spaces left between endoderm and ectoderm. Actually these body spaces are simply remains of the embryonic blastocoele that do not fill in with mesenchyme after the gut forms. These spaces are called **pseudocoelomate** because they are not lined by any embryonic mesoderm layer. Pseudocoelomate animals, in lacking the germ layer mesoderm, have poorly developed musculature or organs that are primarily formed of mesoderm in higher animals.

True coelomate animals originated when mesoderm formation began to occur at the same time the endoderm and ectoderm layers became established embryonically. In describing spiral cleavage, we noted that the cell to be devoted to forming mesoderm could be identified very early. When mesoderm forms as a layer or bundle of tissue, and then splits or expands so that it covers both the gut and the inside of the body, a true coelomic cavity is formed. This type of mesodermal distribution permits the formation of muscular layers around the gut and underlying the ectoderm. It also provides tissue for a closed circulatory system, to augment simple diffusion as a means of supplying nutrition and waste disposal.

Coelom Formation. The method of forming the coelomic cavity in the mesoderm is one of the major differences between the two main animal groups mentioned earlier, the annelid-arthropod line and the echinoderm-chordate line.

The annelid-arthropod groups usually have spiral cleavage in their eggs. During gastrulation, the coelom is usually formed by the splitting of the center of a solid sheet of mesoderm, a method known as **schizocoelous.** The echinoderm-chordate groups, which show radial cleavage in the eggs, generally form the mesoderm and coelomic cavity by outpouchings of the primitive gut endoderm, as we saw in *Amphioxus.* This is the **enterocoelous** method.

Another difference is in the fate of the blastopore after gastrulation. In the annelid-arthropod groups the mouth forms in the blastopore region; such animals are known as Protostomia. In the echinoderm-chordate groups, the blastopore region becomes the anus and a new mouth breaks through at the opposite end of the gut. These animals are called Deuterostomia.

Intermediate Groups. The echinoderm-chordate line may have arisen as an offshoot of the older annelid-arthropod line. Some weight is given to this possibility by the intermediate characteristics seen in phyla called Lophophorates, several groups whose members have a crown of tentacles on a ridge around the terminal mouth. In these groups, the cleavage patterns, method of coelom formation, fate of the blastopore, and other characteristics tend to be mixed between those typical of the two main lines.

Other Characteristics. Other characteristics that are considered in classifying animal groups are: the presence or absence of appendages, the occurrence of segmentation, and the nature of the excretory, respiratory, and skeletal structures. The basic differences that serve to separate large groups of animals give way to progressively smaller differences as the animals are more closely related. The differences between two species of the same genus may be only slight, based on technical details that are not at all apparent to the casual or untrained observer.

A number of tables, charts, and keys have been constructed by authors over the years in attempting to show the relationships of the phyla to one another. The major difficulty is that individual animals or groups persist in straddling the boundaries of such charts. A typical arrangement, using the various criteria just discussed, is presented here in Fig. 112 and in Table IV. It should be noted that most of the

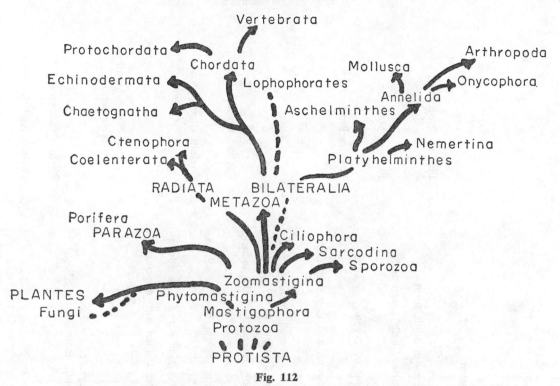

Fig. 112

categories are not valid systematic taxa. The divisions are used simply to emphasize the comparative levels of complexities, and the differences between groups. When using such an outline, one must always progress through each step in the order given without skipping to the smaller separations or minor characteristics.

PHYLUM PROTOZOA

History. The Protozoa are microscopic unicellular or acellular animals that lack tissues or organs. Since they are so small the Protozoa were of course not recognized until the advent of the microscope. Leeuwenhoek referred to them as the "animalcules," but this included all the various kinds of organisms he observed with his lenses without attempting to distinguish between them taxonomically.

The name Infusoria was given the animalcules in the eighteenth century because of the growth of the organisms in jars of water, containing hay or other vegetation, known as infusions. Among the organisms were many that actually belong to higher groups, and it was not until about 1840 that the acellular protozoans were separated from the others. The word Protozoa ("first animal") was coined in 1818, but applied to all Infusoria. Only after the development of the cell theory, in the mid 1800s, did the term Protozoa achieve its present usage.

Morphology. Since the acellular animals do not have cells with which to form organs, specialized areas of the individual perform this task. Such differentiated areas or structures are called organelles, and include cilia or flagella for locomotion, vacuoles for storing food and water, sensory fibrils, and secreted coverings or tests.

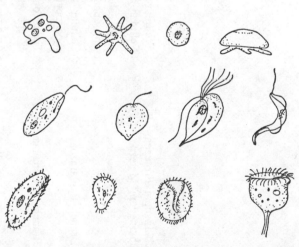

Fig. 113. Protozoans. *Top to Bottom*—Sarcodina, Flagellata, Ciliata

TABLE IV

The Animal Kingdom

Classification	Description	Phylum
I. Subkingdom Protozoa	Acellular animals	Phylum Protozoa
II. Subkingdom Metazoa	Cellular animals; some lose cell boundaries in adult form	
Branch A. Mesozoa	Solid cellular animals with a layer of surface cells plus interior reproductive cells	Phylum Mesozoa
Branch B. Parazoa	Cellular grade, beginning tissue formation; no organs, digestive choanocyte cells	Phylum Porifera
Branch C. Eumetazoa	Tissue or organ grade of construction; mouth and digestive tract except where degenerated	
Grade I. Radiata	Animals with primary radial symmetry, tissue grade, some organs; mesenchyme ecto-mesodermal, mouth and digestive cavity	
Level A.	Symmetry radial or biradial; mouth usually ringed by solid tentacles; stinging cells called nematocysts	Phylum Coelenterata
Level B.	Symmetry biradial; tentacles not encircling mouth, rows of ciliated swimming plates or combs	Phylum Ctenophora
Grade II. Bilateralia	Organ grade construction, bilateral or secondarily radial as adults	
Level A. Acoelomata	Space between body wall and digestive tract filled with mesenchyme; excretory system protonephridia or flame bulbs	
	1. Anus absent	Phylum Platyhelminthes
	2. Anus present	Phylum Nemertina
Level B. Pseudocoelomata	Body cavity present as remains of embryonic blastocoel	
	1. Intestine looped, anus near mouth; both encircled by ciliated solid tentacles	Phylum Entoprocta
	2. Intestine with posterior anus	
	a. Body covering a scleroprotein cuticle	Phylum Aschelminthes
	b. Body wall with fluid-filled lacunae or channels not open to interior or exterior; spiny heads	Phylum Acanthocephala

Level C. Coelomata

True body coelom lined with endo-mesoderm layer; excretory organs protonephridia or metanephridia

1. Lophophorata
   With a lophophore, a circular or crescentic ridge bearing ciliated hollow tentacles around the mouth only
   a. Colonial; no nephridia or circulatory system . . . . Phylum Ectoprocta (Bryozoa)
   b. Solitary; with metanephridia and closed circulatory system
      (1) Wormlike . . . . Phylum Phoronida
      (2) With bivalve shell . . . . Phylum Brachiopoda

2. Without lophophore
   a. Schizocoela or Protostomia. Coelomic cavity formed by splitting of mesodermal layer; blastopore site becomes mouth
      (1) Unsegmented
         (a) Viscera covered by body fold (mantle) which secretes a calcareous shell . . . . Phylum Mollusca
         (b) Body covering a cuticle
            1. Anus posterior, nonretractile proboscis . . . . . Phylum Echiurida
            2. Anus antero-dorsal . . . . Phylum Sipunculida
      (2) Segmented
         (a) No appendages . . . . Phylum Priapulida
         (b) With appendages
            1. Appendages not jointed . . . . Phylum Annelida
            2. Appendages jointed . . . . Phylum Arthropoda
   b. Enterocoela or Deuterostomia. Coelom formed as outpouching of digestive tract to form mesoderm; blastopore site becomes anus
      (1) Secondary radial symmetry; with water-vascular system . Phylum Echinodermata
      (2) Bilateral symmetry throughout life
         (a) Without gill slits or endoskeleton . . . . Phylum Chaetognatha
         (b) With gill slits
            1. Without notochord . . . . Phylum Hemichordata
            2. With notochord, endoskeleton . . . . Phylum Chordata

The shapes may be round, ovoid, vase-shaped, or similar variations of spherical symmetry. Many protozoans are able to move by extending protoplasmic pseudopodia (false feet) and oozing along shapelessly, but some have pseudopodia that are stellate, fixed, and regular in formation. Flagella often wave about in the manner of a boat propeller, pulling the animal forward in a spiral path. Cilia beat in coordinated waves to provide motion.

Physiology. Protozoans show a wide range of feeding methods. Some are holophytic (autotrophic), employing a process identical to photosynthesis in plants. They synthesize carbohydrates from water and carbon dioxide by means of chromoplasts of various colors and sunlight. Proteins are built by combining nitrogen salts with carbohydrates. Some of the chlorophyll-bearing flagellates are able to live in a completely inorganic solution if it contains ammonium nitrate. Others need only the addition of organic carbon or nitrogen to survive; no flagellates without chlorophyll can do this.

"Saprophytic" and "saprozoic" are terms applied respectively to the plantlike and animal-like protozoans that do not ingest food but absorb amino acids or peptones from the medium in which they live. Both the free-living colorless flagellates and the parasitic forms may feed in this way. A few show what may be considered exotic tastes for acetic or butyric acid and ammonium salts.

Mixotropic or **mesotropic** nutrition is a combination of holophytic and saprophytic modes, perhaps reflecting a transition from one method to another. Holozoic nutrition represents the true animal mode of feeding, with solid food being ingested. The protozoans may feed on bacteria, yeasts, algae, other protozoans, and on small metazoans. Most of the free-living protozoans feed in the holozoic mode. A few special environmental feeders are recognized; protozoans that live in fecal material are said to be **coprozoic** and those that live in bottom slime of foul waters are called **sapropelic.**

Respiration is accomplished by diffusion of oxygen through the cell membrane. Those that have adapted to living without oxygen, beneath the surface of a medium, obtain their energy through the anaerobic glycolytic cycle described previously.

Excretion also occurs by diffusion through the cell membrane; the waste product is ammonia. Many protozoans have a vacuole into which excess water in the cell is channeled. This vacuole then contracts, expelling the water; hence it is known as the contractile vacuole. Other vacuoles are present in many species; these represent food that has been engulfed and circulated through the cytoplasm as digestion takes place. Contractile vacuoles are commonly seen in phytomonads, euglenoids, and fresh-water sarcodinans.

Some sensory perception is shown by the protozoans; the phytoflagellates often bear light-sensitive spots or **stigma,** and they exhibit attraction to moderate amounts of light but move away from more intense light. Almost all show varying degrees of avoidance behavior; certain species show the ability to detect their favorite prey or their predators, and take appropriate action to approach or avoid.

Reproduction. Protozoans show a wide range of reproductive methods. The sexual method is rare among the flagellates, and usually consists of the merging of isogametes into a single individual (syngamy). After an encysting period, the single individual divides into two, four, or many new individuals. Multiple fission is not uncommon; a number of individuals may merge into a mass called a **palmella stage** which later breaks up into many spores or individuals. Binary fission in flagellates is usually longitudinal, although some dinoflagellates divide diagonally. Ciliates divide by transverse fission. They do not have free sexual gametes, but may merge or may exchange nuclear material.

Organelles. Organelles include cilia and flagella. At one time great distinction was made between the two structures, but the electron microscope has shown that the internal composition is identical. Both are formed by two central cores of fibrils and a circle of nine double fibrils. The outer sheath is continuous with the cell membrane. Most flagellates and ciliates show an anterior end, and the flagella are usually attached there. Cilia tend to be either uniformly distributed over the body or concentrated near the anterior end. Flagella are usually much longer than cilia, and fewer in number. The dividing line, however, can be difficult to determine. When do long cilia become short flagella? Flagella often show parabasal bodies (blepharoplasts), which appear to anchor and guide the flagella. Some also show a rod, the **axostyle,** which runs through the length of the organism. Sometimes a membrane extends along the surface and its undulations aid in locomotion. Ciliates have a system of fibrils that connect the cilia beneath the surface, and are apparently neuromuscular, controlling and co-ordinating the beat of the cilia. Parabasal granules are linked to the fibrils.

The flagellates are considered to be the stem group from which the metazoans (multicellular animals) developed. Flagellated sperm cells are common in plants and almost universal in animals. The parabasal bodies to which the flagella attach appear to act as centrioles during cell division. Flagellates show almost all the possible modes of nutrition, and also include colonial forms. *Volvox,* a colonial form mentioned earlier, is a phytoflagellate that has become organized with specialized anterior sensory functions and posterior reproductive functions. The offspring grow in the hollow center of the colony, which swims by co-ordinating the flagellar beat of the individuals. A zooflagellate, *Proterospongia,* is a cluster of cells, each of which has a protoplasmic collar surrounding a single flagellum. Such cells, called **choanocytes,** are seen in sponges.

Ecology. Protozoans are almost universally distributed. Many are free-living, but many have become adapted to life in relation to some multicellular organism. Some are **symbiotic,** which means they live together as a matter of convenience; some are **commensal,** eating together; and some are **parasitic,** which means that one lives at the expense of the other.

Perhaps the most interesting commensal relationship is that of the flagellates that inhabit the termite gut. Termites feed only on wood, yet they are unable to break the bonds of cellulose in digestion. The flagellates are able to break these bonds, making the food available to protozoans and termites alike. Without this fauna in the gut the termite would starve to death.

Parasites. Other protozoans are not so convenient for the host to harbor. Amoebic dysentery is a common disease in countries that lack sewage disposal and water purification. The trypanosomes, which are zooflagellates, cause African sleeping sickness, and Chaga's disease in South America. Trypanosomes have a flagellum plus an undulating membrane, and require intermediate hosts for separate stages in their life cycles.

Elaborate life cycles are often typical of parasitic animals, enabling them to be transported from one victim to another. It is not efficient from the parasite's point of view to kill a host, but only to use it. And if the host dies without the parasite having infected another host, that is not efficient either. Insects that bite and suck are common carriers of parasites. Usually the stage or stages that are found within the insect differ from the forms seen in the primary host; they may be in a dormant state or they may multiply sexually or undergo rapid cell division or fragmentation (sporulation).

African sleeping sickness is caused by *Trypanosoma gambesiense* or *T. rhodesiense,* which are carried by different species of the tsetse fly. One form of sleeping sickness is long and debilitating while another is acute and often fatal within a short time. The organisms settle in the lymph nodes and bloodstream, causing muscle soreness, fever, swelling, and sleep. A related genus, *Leishmania,* causes such diseases as oriental sore and kala-azar, a deadly disease characterized by swollen viscera and spleen. These are carried by fleas. Chaga's disease is carried by *Triatoma,* the cone-nosed bug.

Another zoomast group, the trichomonads, contains species that may cause irritation and discharges in the mouth, vagina, and intestine. Many species that have not been demonstrated to cause disease are present in animals also.

The Sporozoa are a group of parasites with highly specialized life cycles. It was once thought that they lacked locomotor organelles, but some have been observed to have flagella or pseudopodia at some phase of their life cycles. Thus their origins are open to question. All are minute and form spores which may or may not have a small polar capsule containing an eversible barb. Probably the most economically significant are the haemosporidians, *Plasmodium vivax,* *P. falciparum,* and *P. malariae,* which cause the three common types of human malaria.

Malaria, meaning "bad air," which the citizens used to blame for their fevers, was until recently one of the scourges of mankind. It is carried by several species of *Anopheles* mosquito. One can trace the life cycle by beginning with the fertilized zygote (oökinete) in the stomach of a mosquito that has fed on a malaria carrier. The oökinete travels in wormlike fashion into the gut lining, where it forms a cyst (oöcyst). There it divides many times inside the cyst, forming many sporozoites. When the cysts burst, the sporozoites pass through the tissues of the host mosquito until they come to the salivary glands. When the mosquito feeds on an animal, it first releases a small amount of saliva to soften the skin, and in this way the sporozoites enter the bloodstream of the human host. The sporozoites then are carried to the liver, where they are engulfed by liver cells. They have become trophozoites, and inside the liver cells they divide many times to become schizonts. These mature into a new form, the cryptozoites, which burst from the liver cells and enter the bloodstream. There they invade the red blood cells where they reproduce by multiple fission to become merozoites. When the red blood cells rupture, the merozoites re-enter the bloodstream where some invade new red blood cells while others become male and female gametes. It is at this stage that they may be sucked up by another mosquito when it sucks blood to feed, and fertilization takes place in the mosquito gut to begin the cycle again.

The fever of malaria occurs in cycles, varying with the species; the cycles correspond to the intervals when the red blood cells rupture, releasing the toxic metabolic products of the merozoites, as well as the organisms themselves, into the bloodstream. Control

TABLE V

Classification of Phylum Protozoa

Subphylum Plasmodroma–locomotion by flagella or pseudopodia;
monomorphic nuclei, sexual
reproduction by syngamy (fusion).

Class Flagellata (Mastigophora)–flagellates
Subclass Phytomastigina (plantlike)

Order Chrysomonadina	–1, 2 flagella; yellow, brown chromoplasts
Cryptomonadina	–2 flagella; colored or not
Euglenoidina	–1, 2+ flagella; green, brown chromoplasts or not; store carbohydrates
Chloromonadina	–2 flagella; chloroplasts; storage product fat
Dinoflagellata	–1, 2 flagella; tests with grooves
Phytomonadina	–1–8 flagella; green; some colonial

Subclass Zoomastigina (animal-like)

Order Protomonadina	–1–3 flagella; colorless; have parabasal body
Polymastigina	–2–6 flagella, undulating membrane; parabasal body, axostyle; many parasitic
Hypermastigina	–many flagella; parabasal body; commensal in termite gut
Rhizomastigina	–1 flagellum; amoeboid, colorless

Class Sarcodina–with pseudopodia
Subclass Rhizopoda–lobose pseudopodia without axial rods

Order Amoebina (Lobosa)	–naked, asymmetrical; most free-living
Testacida	–fresh water, in single-chambered shell (test) with one opening
Foraminifera	–most marine, multichambered test, many openings for pseudopodia

Subclass Actinopoda –with axopods (stiffening rods in pseudopodia)

Order Radiolaria	–marine; lattice-like skeleton
Heliozoa	–fresh water; spherical with stiff radiate axopods
Helioflagellata	–with axopods, one or more flagella

Class Sporozoa–parasitic; reproduction by multiple fission
Subclass Telosporida

Order Gregarinida	–the gregarines; wormlike, extracellular in invertebrates
Coccidia	–coccidians; intracellular in blood, gut; cause coccidiosis disease
Haemosporidia	–in blood of vertebrates; blood-sucking intermediate host; cause malaria, cattle fever

Subclass Cnidosporidia –with capsules in spores having reversible barbs like nematocysts

Order Myxosporidia	–fish parasites; cause tumors
Actinomyxidia	–worm parasites, 3-rayed spores
Microsporidia	–arthropod, fish parasites; cause silkworm disease

Subclass Sarcosporidia — muscle cysts in mammals
Haplosporidia — uncertain affinities
Subphylum Ciliophora—locomotion by cilia; sexual reproduction by conjugation (exchange)

Class Ciliata—cilia for locomotion and food-catching
Subclass Protociliata
Order Opalinida — parasites in frog, toad intestine; no mouth or contractile vacuole; monomorphic nuclei
Subclass Euciliata — polymorphic nuclei; mouth, contractile vacuoles
Order Holotricha — with simple, uniform ciliation; trichocysts (minute dischargeable rods)
Spirotricha — fused cilia around mouth form organelle; peristome area large
Conotricha — epizoic on amphipods; vase-shaped, ciliated funnel peristome
Peritricha — vase-shaped with attaching stalk; cilia around peristome

Class Suctoria—tentacles replace mouth and cilia; attached to substratum

---

in the United States has been based on eliminating mosquito-breeding areas such as swamps. Even an ornamental bird bath or fish pond may furnish ample water for the mosquito larvae, which are aquatic. Drug therapy is difficult in humans because the organisms are protected inside the liver cells and blood cells for much of the life cycle. Reptiles, birds, apes, and monkeys may also contract malaria.

Ciliates are fairly common parasites or commensals in a number of vertebrates and invertebrates, but only one, *Balantidium coli,* is found in man. It may cause diarrhea and intestinal pain.

PHYLUM MESOZOA

The mesozoans are a small group of obscure parasites whose origins and relationships are unknown. During the sexually reproductive stage the body consists of center reproductive cells surrounded by a single layer of somatic cells.

History. Little is known of the mesozoans other than that they have complex reproductive cycles. Some authorities believe them to be degenerate flatworms, while others believe that they probably arose independently from a multinucleate ciliate in a line of development which has progressed no further. It would thus be an offshoot of the evolutionary tree, just as the sponges are thought to be by many authorities.

TABLE VI

Classification of Phylum Mesozoa

Class Mesozoa
    Order Dicyemida—parasitic in squid, octopus nephridia
    Orthonectida—parasitic in body cavity of various worms, clam

Physiology. Little is known of the physiology of the mesozoans. The reproductive cycle of the dicyemids is complex. The wormlike sexual form, a nematogen, produces other nematogens from the large reproductive cell and then produces an egg. The egg develops into a larva, which escapes presumably to an intermediate host where it passes an asexual reproductive phase. The orthonectids are free-swimming during the sexual phase, and the egg develops in the female. The larva then invades the host where it forms a plasmodium, a multinucleate mass that undergoes fragmentation. Males and females develop from some of these fragments and leave the host. They do not appear to be injurious to the host animals. Authorities are not in agreement on the details of the life cycles.

PHYLUM PORIFERA

History. The sponges are the most primitive of the metazoan multicellular animals. Apparently the mutations that led to multicellularity in the sponges did not include provisions for neuromuscular co-ordination, so the group progressed little throughout its long period of existence. Since the sponges are sessile, and most are attached, it is not surprising that they were thought to be plants by the ancient naturalists. In the eighteenth century the water currents produced by flagellated cells were discovered, and the animal nature of the group gradually was accepted. At first they were classed with the other radiate animals, but R. E. Grant recognized their very simple level of organization and named them the Porifera, pore-bearers.

Morphology. The smaller, simple sponges are usually vase-shaped, with pores leading into a central cavity, but most of the sponges are irregular, encrusting, or branching masses. These animals are of a cellular level of organization, without tissues or organs and lacking a mouth and digestive cavity. The body is provided with pores and canals through which water enters and circulates. The in-current pores (ostia) are numerous; exit of the water from the central cavity is through one or a few large openings (oscula). Interior body cavities are lined with flagellated collar cells, choanocytes, which perform digestion intracellularly (within each cell). Most sponges have calcareous or siliceous spicules embedded in the body walls. There is a high degree of independence to the individual cells of sponges. Large mesenchymal cells in the space between epidermis and flagellated cells perform many tasks—secreting spicules, transporting nutriments from the flagellated choanocyte cells, and probably forming the gametes.

Classification. The classification of the sponges is based almost entirely on the nature of the spicules. Some are formed of calcium; some are siliceous; others have horny fibers in place of spicules. The shape of the spicules—the number of radii or axes—determines the classification.

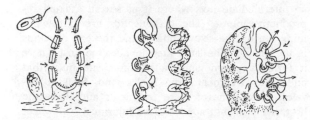

Fig. 114. Sponge Types. *Left to Right*—Ascon, Sycon, Leucon

TABLE VII

Classification of Phylum Porifera

Class Calcarea	—skeleton composed of separate calcareous spicules
Order Homocoela	—ascon sponges (simplest body type)
Heterocoela	—syconoid or leuconoid body types
Class Hexactinellida	—glass sponges, with six-rayed siliceous spicules
Demospongiae	—horny sponges, some also with siliceous spicules, nontriaxon; leuconoid body
Order Monaxonida	—with monaxon spicules
Tetractinellida	—with tetraxon spicules
Keratosa	—horny fibers

Physiology. Water currents produced by the beating of the flagellated cells bring oxygen and food to the cells of the sponge and carry away the wastes. A sponge 10 centimeters high has been measured as "pumping" 22.5 liters of water through it in one day. The collar cells trap particles of detritus and planktonic organisms, and digestion is within these cells or in the amoebocytic cells that move through the interior. Authorities disagree on whether the choanocytes or the amoebocytes produce the gametes, but egg and sperm are "nursed" by amoebocytes until they are released into the water.

Embryology. The eggs may be retained in the mesenchyme, and the sperm, carried in by the water currents, are transported into the interior by choanocytes for fertilization of the eggs. Cleavage produces two tiers of eight cells each; one becomes epidermis and the other becomes flagellated cells. The flagellated cells grow rapidly until they surround the epidermal cells; then the entire mass turns inside out so that the flagellated cells are inside and will become choanocytes. Different species show other peculiar changes that are the reverse of the usual gastrulation seen in higher groups.

Ecology. The largest keratose sponges, up to barrel size, are found in deep tropical waters. Before the advent of plastic sponges there were areas of Greece whose entire economy was based on diving for sponges. Many Greeks migrated to western Florida and established the sponge industry there. Other sponges, not so large and important, are found in many marine areas, and one family is found in fresh water. One marine family is able to bore into calcareous structures such as mollusk shells and coral, which can be economically damaging.

PHYLUM COELENTERATA (CNIDARIA)

History. Coelenterata include over nine thousand living species and have an extensive fossil record because of the preservation of the calcareous coral formations. Most species are attached during at least part of the life cycle and many form branching colonies, so the coelenterates were long considered to be plants. Part of the life cycle also may be spent as a free-swimming jellyfish. These forms were not recognized as being related to the attached forms for many years, and sometimes the two stages were given different species names.

The name Coelenterata, meaning "digestive cavity," was applied to a large assemblage of animals at various times, but is now restricted to those tissue-grade animals that bear stinging cells called **cnidoblasts**—hence the name Cnidaria is sometimes given the group. The cnidoblast cell everts a barbed tubule called a **nematocyst** through which a toxic fluid may be injected into prey or predator.

Morphology. Coelenterates are radially symmetrical, although some are modified to biradial or bilateral form. Two body types occur: the polyp or hydroid, and the medusa, shown in Fig. 115. The polyp consists of a base anchored in the substrate, plus a hollow stalk, and one or more branched oral ends, depending upon whether it is a solitary or colonial form. Tentacles surround the oral opening.

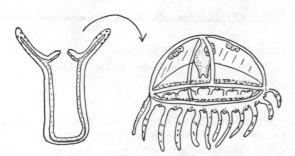

Fig. 115. Coelenterate Body Types. *Left*–Sessile Polyp. *Right*–Free-swimming Medusa

The medusa body form consists of a solitary gelatinous umbrella, usually with tentacles suspended from the edge. The mouth and gastric cavity are suspended in the center like an umbrella handle. From the gastric cavity four or more radial canals extend across the umbrella to a ring canal around the margin. A thin muscular membrane, the **velum**, may lie just inside the margin behind the tentacles. The presence or absence of the velum distinguishes the body types. The medusa swims by contraction of the umbrella, which expels water and provides a primitive jet propulsion.

TABLE VIII

Classification of Phylum Coelenterata

Class Hydrozoa	–hydroids, and medusae with velum
Order Hydroida	–hydroid stage dominant
Hydrocorallina	–hydroid stage dominant
Trachylina	–medusa stage with velum dominant
Siphonophora	–floating colonies of both polyp and medusa attached
Class Scyphozoa	–true medusa without velum (the jellyfish); no polyp stage
Class Anthozoa	–polypoid, no medusa stage (anemones and corals)
Subclass Alycyonaria (Octocorallia)	
Order Stolonifera	
Telestacea	
Alcyonacea	–(soft corals)
Gorgonacea	–(gorgonian corals, horny corals, sea fans)
Pennatulacea	–(sea pansies, sea pens)
Coenothecalia	–(blue corals)
Subclass Zoantharia (Hexacorallia)	
Order Actiniaria	–(sea anemones)
Madreporaria	–(stony [true] corals)
Antipatharia	–(black corals)

Physiology. Coelenterates are composed essentially of an epidermis and a gastrodermis separated by a gelatinous mesoglea. The epidermal layer contains epithelio-muscular cells, mucus cells, sensory and nerve cells, blastic interstitial cells capable of forming any of the foregoing, and the unique cnidoblast cells. Each cnidoblast contains a tubule nematocyst that may be discharged; when a change in the permeability of the cell wall capsule occurs, water rushes in and the tubule everts. The individual tubule end may be modified so that it is threadlike and can entangle prey, is sticky and helps fasten the animal to the substrate, or is barbed and a toxin can be injected through it. The toxins of the smaller coelenterates are not felt by man, but the large jellyfish can inflict severe, even paralyzing, injury.

The gastrodermis is capable of extracellular secretion, which partially digests proteins brought into a cavity by the tentacles. Digestion then continues on an intracellular level. Coelenterates are apparently not able to digest starches. Fat and glycogen are the storage products, which are passed among the cells by diffusion. Indigestible materials must be ejected through the mouth, since the cavity has only one opening. In an aquarium an anemone about six inches in diameter captured and ingested an octopus larger than itself, which gives some indication of the muscular stretch and digestive capabilities of these animals.

There are no circulatory or excretory systems, both functions being accomplished by diffusion. A simple nerve net composed of bipolar cells occurs in the simpler forms, but some of the more advanced forms show the development of synaptic connections between sensory and motor cells.

Embryology. Coelenterates typically reproduce both sexually and asexually, with gametes being formed by interstitial cells only at certain seasons. The gametes collect into temporary gonads, which are said not to be true organs. Hydroids bud throughout most of the year. The life cycles often are complex; apparently the original body type was the medusa, which was free-swimming and reproduced sexually. Gradually the ciliated larvae that were produced began to undergo a prolonged stage of settling and attaching to the substrate, where they grew into the polypoid body form. Tiny medusae were formed on specialized polyps of the colonial growth, which were released at maturity to become the free-swimming, sexual stage. Some forms do not release the medusae, which give off gametes while remaining attached to the polyp.

Some of the hydrozoans presently show both stages of this type of life cycle but most coelenterates have become specialized in the direction of one or the other body plan for most or all of their life span. Thus we see the *Hydra* and anemones, which never go through a medusa stage, and various medusae that have eliminated or never had a polypoid stage.

The fertilized egg undergoes complete cleavage and forms a hollow blastula. Gastrulation is accomplished by in-wandering of cells until a stereogastrula is formed. This then elongates into a radially symmetrical, ciliated larva called the **planula.**

PHYLUM CTENOPHORA

The Ctenophora are marine planktonic animals somewhat resembling coelenterates, from which they no doubt evolved. There are only about eighty species, although such common types as the so-called sea walnut may at times be found drifting in large numbers. The basic body type appears to be a modification of the medusoid form. It is biradial, globose, with eight rows of fused cilia called **comb rows** extending from the oral to the aboral end. A pair of sheathed tentacles extend like dangling arms from the sides of the body. The mouth hangs down and leads to a digestive cavity that is subdivided so that a tube lies beneath each comb row. The space between the epidermis and gastrodermis is filled with mesogleal jelly.

Ctenophores are of a higher structural level than coelenterates. Muscle cells are found in the mesoglea, in some cases grouped into fibers. The digestive tract actually has an anal opening at the aboral end in some species, although this characteristic is not seen in the bilateral phyla until a much higher group is examined. Sensory spots are found on coelenterate medusae, and there is a sensory dome that contains a statolith lying atop the aboral end. A statolith is a pebble-like concretion that rests on tactile hairs; when the animal changes position the pressure of the shifting stone relays a sensory impression to the comb rows which undulate to provide motion.

TABLE IX

Classification of Phylum Ctenophora

Class Tentaculata —with tentacles (*Pleurobrachia*)
      Nuda   —without tentacles (*Beroë*)

Physiology. Only one species of ctenophore bears nematocysts. All the others have adhesive cells, called **colloblasts**, on the tentacles, with which they capture prey.

Although respiration and excretion presumably take place by diffusion, there are clusters of cells called **rosette cells** in the digestive tract that seem to be osmoregulatory in nature. Excretory cells also do not occur in the lowest of the bilateral groups. Modification of the basic ctenophore biradial body plan seems to be in the direction of lengthening the tentacular plane and flattening the oral-aboral plane.

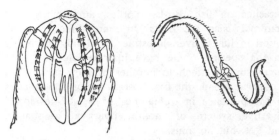

Fig. 116. Ctenophores

One such example is found in tropical seas—the Venus's-girdle, a long streamer that is luminescent, as are many of the ctenophores.

Embryology. All species are hermaphroditic. Gonads arise in the meridianal canals and the products exit either through the epidermis or through the canals. Cleavage is total, forming a stereoblastula of micromeres and macromeres. The micromeres grow down over the macromeres, which invaginate to form the gastrula. The mesenchyme appears to be derived from the ectoderm, as it is in the coelenterates.

Ecology. The ctenophores form at times a considerable part of the plankton in the sea. Their principal economic significance is that they feed particularly on oyster larvae, and can destroy a large part of the crop in a short time.

Ctenophores are considered to be specialized, and not in the main line of evolution of the Bilateria. However, a number of factors, such as the existence of rosette cells and anus, give rise to question. One species that has become modified to a creeping mode of life strongly resembles one of the higher flatworms with a much-branched gut. This is regarded by some as parallelism in evolution rather than direct evolution, since both forms are among the most highly specialized of their respective phyla.

PHYLUM PLATYHELMINTHES

The Platyhelminthes are the most primitive group of the bilateral metazoans. Within the group there are organisms that are little advanced over some of the most advanced ciliated protozoans, as well as some that are highly specialized. The most primitive flatworms (Order Acoela) are little more than a ciliated epidermis covering a mass of cells, some of which are digestive in function. An opening in the ventral epidermis allows the digestive cells to come in contact with food. For some years certain author-

ities believed that these worms represented forms degenerated from more complex types, but this is now considered highly unlikely.

Origins of the Bilateralia. The question of how the bilateral groups arose from the primitive radiate groups has been an intriguing one to scientists and several theories have been advanced. One theory supposes that the multinuclear ciliates, by forming cell divisions, gave rise to a ciliated larval type such as the planula seen in the coelenterates. If such a larval type failed to attach to the substrate and change into the adult form, it would approach the basic body plan of the acoel flatworms. If the animal adopted a creeping existence, dorsoventral flattening would gradually take place and the nerve centers would shift to the anterior portion.

One difficulty in postulating a ciliate ancestry is the almost universal existence of flagellated sperm. Nuclear patterns also tend to link the flagellates to the metazoans. A planuloid larval body plan is considered to be basic to the higher groups, giving rise to the medusa and polyp in coelenterates, and to the acoel flatworms. Flagellated gametes are postulated. Actually, there is no good reason for assuming that only one group gave rise to the entire metazoan complex; it is possible that different protozoan groups resulted in various metazoan groups, but because there is no fossil record no accurate answer to the question of metazoan origins can really be given.

Morphology. Two of the three classes of Platyhelminthes are entirely parasitic: the Trematoda and Cestoda, which have become highly specialized in adapting to parasitism. Most members of the class Turbellaria are free-living and are thus thought to be more typical of the phylum.

The Turbellaria are flattened and ovoid or elongated in shape. Head projections are often present in the form of tentacles or auricles (earlike lobes). Eyes are also quite common. The outer body surface is a ciliated epidermis and the body is a solid mesenchymal mass lacking cavities; it is therefore acoelomate in structure. Turbellarians have numerous gland cells in the epidermis and in the mesenchyme which secrete a mucus slime over the entire body. This mucus aids both in movement and feeding. Some secretory areas are further modified with muscle fibers to form adhesive discs. Peculiar rodlike concretions, called rhabdites, can be seen histologically in the epidermis; these can be ejected from the surface and apparently dissolve into a sticky mass to aid in catching food. There are no circulatory or respiratory systems, but excretory structures called flame cells are present. A tuft of cilia forms a flame-shaped organelle that gives the cell its name.

The orders of the turbellarians are based primarily on the nature of the digestive tract. This structure progresses from a simple sac in some groups to a three-branched structure with many diverticula in the Tricladida, to a multibranched structure in the Polycladida.

Cestoda. The Cestoda (Fig. 117) are parasitic tapeworms that inhabit the gut of many vertebrate animals as the primary host. As is typical in parasites, intermediate hosts are usually required for completing the complex life cycle and transferring the parasite from host to host. Although some cestodes are not segmented, most consist of a head (scolex), which is armed with hooks or suckers to anchor into the intestinal wall, and a series of identical segments attached to the scolex like boxcars to a train engine. Each segment (proglottid) is little more than a reproductive factory, containing both male and female systems. The oldest segments are at the end. The testes mature first in the younger segments and can fertilize the eggs that are produced in the older segments after the testes degenerate. Mature proglottids may contain thousands of eggs that may be discharged from the genital pore, or the segments themselves will break off and be passed from the host in the feces. The only connections between segments are excretory tubules and nerve cords along either edge.

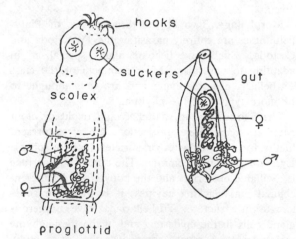

Fig. 117. Platyhelminthes. *Left*–Cestoda. *Right*–Trematoda

Trematoda. The trematodes, or flukes, differ from the cestodes in a number of ways. They are generally flattened, ovoid, and have one or two suckers which may be modified by hooks and bars for attachment. Trematodes parasitize a wider range of animals than cestodes; some are found on or in invertebrates although the largest number are ecto-

parasites on fish. Birds, reptiles, amphibians, and mammals also are hosts. In man, flukes may be found in lungs, liver, intestine, or blood, depending on the species. Some have intermediate hosts but many simply attach to another primary host when the larvae escape the first host. A major difference is in the presence in trematodes of a well-developed digestive system of mouth, pharynx, esophagus, and gut, but no anus.

TABLE X

Classification of Phylum Platyhelminthes

Class Turbellaria	–free-living
Order Acoela	–no gut (*Polychoerus*)
Rhabdocoela	–simple sac gut (*Anoplodiera*)
Alloecoela	–simple gut with diverticula
Tricladida	–3-branched gut (Planarians)
Polycladida	–many branched gut, marine (*Leptoplana*)
Class Trematoda	–parasitic flukes (*Opisthorchis*)
Cestoda	–parasitic tapeworms (*Taenia*)

Physiology. Except for the primitive acoel flatworms, the organ level of function and development has been reached in Platyhelminthes. Obviously, the nutritional activities will vary among the groups because of the parasitism occurring in so many. The free-living planarians, fresh-water inhabitants of many of the North American streams, are carnivorous and feed by ejecting an eversible pharynx which attaches to the prey. Digestion is partially extracellular, being completed by phagocytic cells. Fat and starch apparently can be digested and fat is the storage product.

In recent years much speculation has arisen over some experiments that indicated the planarians could be trained by shock conditioning, and that "memory" (the ability to become retrained in a shorter time after a time lapse) could actually be transferred to another individual by cannibalism. This is very significant, because it has been thought that only the much higher animals are capable of memory or training. The popularity of RNA studies during the same period has led some investigators to conclude that RNA molecules provided memory storage. As is often the case, first results seemed more positive than later ones; it appears now that planarians will retrain more quickly under any one of various stimuli other than the cannibalism of trained planarians, stimuli that have no possible connection with a memory mechanism.

Embryology. The patterns of spiral cleavage no doubt have their origins in the Turbellaria. However, these patterns are highly modified in modern flatworms, partly because of peculiar yolk distribution patterns. The acoel flatworms still retain the primitive reproductive method of producing gametes in the mesenchymal cells; gonads are only transitory, if they form at all. In the higher groups, the reproductive systems can be very complex, with accessory sacs for yolk or shell production and storage. All flatworms are hermaphroditic, but they do not self-fertilize. Copulatory organs enable each animal to impregnate the other when two come together to mate. Some species produce "summer eggs" with thin shells, and "winter eggs" with heavy shells for protection and overwintering.

Flatworms often possess remarkable powers of regeneration. *Planaria* cut into many pieces will regrow an entire whole animal from each piece. Sometimes they break into pieces spontaneously, a process that could certainly be classed as asexual reproduction.

Life Cycles. The "eggs" of the parasitic forms are often developed to the larval stage, although they are still within capsules when they appear in the feces of animals. The spread of tapeworms is due to improper disposal of wastes, so that the eggs are eaten by the intermediate hosts, which may be various insects, cattle, hogs, or fish, in the cases of human tapeworms. The larvae are released when the capsule is digested in the animal gut, and they burrow into muscle tissues where they encyst. When humans eat raw or poorly cooked fish, pork, or beef, the encysted stage (cystocercus) is activated and attaches to the intestinal wall. It is obvious that in countries where waste disposal is poor and raw meat is eaten, tapeworms will flourish.

The life cycles of trematodes are more complex than those of the tapeworms, involving at least two intermediate hosts. Since trematodes inhabit a variety of body cavities, the eggs reach the outside through feces in some species and through the urine or sputum in others. Almost all inhabit some type of snail as the first intermediate host; either the eggs are eaten when ground snails feed on feces, or they hatch into free-swimming miricidia which burrow into the foot of an aquatic snail. Inside the snail each miricidium encysts and then divides to form many offspring, called rediae. In turn, these reproduce to form cercariae and are released from the snail into water where the cercariae may burrow directly into the bare feet of waders, or into the bodies of fish or mollusks, or they may infest watercress

and water chestnut plants. When raw fish, mollusks, or plants are eaten, the flukes are ingested.

Some species remain attached to the intestinal wall, but others burrow through the wall and migrate through the tissues to the lungs or liver. Other species, of the genus *Schistosoma*, which burrow into the feet, penetrate the blood vessels and cause serious damage by blocking them.

One large liver fluke, *Opisthorchis sinensis,* is a great problem in the Orient. Water chestnuts and fish are often raised in concrete-lined ponds, where snails by the hundreds feast on the algae that grow in the pond walls. The people wade barefoot in the ponds to harvest the watercress and chestnuts, cook their vegetables and fish only little or not at all, and fertilize by dumping night soil (feces) into the ponds. This combination of events allows the spread of *Opisthorchis* in fish as well as several other species on other hosts. One break in the cycle would virtually eliminate the diseases thus caused. Sanitary feces disposal, not wading in infected waters, or proper cooking of food would help greatly. In India, natives use "step wells" where they must stand in the water to fill jars, thus spreading many infections.

PHYLUM NEMERTINA

There are some five hundred species of nemertean worms, which are of solid, acoelomate construction. Because many are long, flat, colorful, and smooth they are known as ribbon worms, although they are also called proboscis worms, after the unusual anterior projection. The long proboscis has a pore and tubule that are usually separate from the digestive tract; the proboscis can be everted from the tubule to snare prey.

Although the nemertean group is small, it is significant because of the many features that appear for the first time in the phylogenetic sequence. Nemerteans are acoelomate and show their flatworm ancestry with a body filled with mesenchyme, a ciliated glandular epidermis, and rhabdites in the proboscis of a few species. The proboscis lies in a coelomoid space (rhyncocoel) which does not appear to be homologous to the true coelom of higher animals. For the first time, this group shows a one-way gut with anus, a simple, closed circulatory system, and a dorsal nerve cord in addition to ventrolateral cords and a "brain." Protonephridia assume an excretory function instead of being purely osmoregulatory, as has been the case in previous groups.

The circulatory system is composed of paired lateral vessels, a median dorsal vessel, and cross connections. Corpuscles bearing red, orange, yellow, or green pigments circulate.

Reproduction may be sexual and dioecious with gametes originating in mesenchymal cells that aggregate to form numerous gonadal sacs. At maturity, ducts also form to extrude the gametes. There are no copulatory structures, but male and female may share a burrow in order to bring sperm and egg together. Asexual reproduction by fragmentation is common.

TABLE XI

Classification of Phylum Nemertina

Phylum Nemertina (=Rhynchocoela, Nemertea)	
Subclass Anopla	–unarmed proboscis
Order Paleonemertini	–most primitive (*Tubulanus*)
Heteronemertini	–(*Cerebratulus*)
Subclass Enopla	–armed proboscis
Order Hoplonemertini	–littoral (*Amphiporus*) fresh-water (*Prostoma*) terrestrial (*Geonemertes*)
Bdellonemertini	–commensal in clams

Nemerteans usually are marine bottom-dwellers that live in and around shells, worm tubules, stones, and algae or form burrows in bottom mud. There is one fresh-water genus and one that is terrestrial in the tropics. As in the flatworms, the tropical terrestrial forms live where there is sufficient rainfall to furnish a very moist environment.

Like flatworms, nemerteans are carnivorous, but they are capable of digesting carbohydrates and fats. Fats are stored in the mesenchyme, and the animals can undergo long periods of starvation while the body becomes greatly reduced. Annelid worms are the primary source of food, but other small invertebrates also are eaten.

PHYLUM ASCHELMINTHES

The Phylum Aschelminthes is perhaps the least stable in concept of all the phyla. It is composed of several classes that differ considerably from one another, and whose relationships are obscure. The chief similarities lie in the fact that all are pseudocoelomate, most possess a complete one-way gut with pharynx region, and cleavage is markedly determinate. A number of authorities prefer to elevate the various classes to phylum status. The ancestry of the groups is questionable; the origins may have been polyphyletic (from several groups) or from a single acoelomate group. Certainly they are all very old in terms of their evolution.

TABLE XII

Classification of Phylum Aschelminthes

Class	Rotifera
	Gastrotricha
	Kinorhyncha
	Nematoda
	Nematomorpha

Because of the differences between the classes each will be discussed separately rather than attempting generalizations that cover all members. The rotifers are frequently seen in fresh-water pond cultures and so were included in the old Infusoria of Linnaeus. For some time they were called Phylum Trochelminthes because they had a superficial resemblance to the trochophore larva, a characteristic of the annelid-arthropod line.

Rotifera. The rotifers are very small, in the size range of protozoans in most cases. They are sometimes called "wheel animalcules" because of a ciliated corona around the anterior end that appears to be rotating when the cilia are beating rapidly. The body is covered by a noncellular cuticle secreted by a syncytial epidermis, and is composed of a head or corona region, a trunk containing the viscera, and a foot sometimes bearing one or two toes. The movements of the animals may be ciliary, or they may jump about by muscular contractions that resemble a child on a pogo stick.

The digestive tract is differentiated into a pharynx that leads to a unique grinding structure called the **mastax.** Esophagus, stomach, and intestine lie in order below the mastax; there is an anal or cloacal opening. The excretory system is composed of two lateral nephridia with flame cells, which is typical of the phylum. The nephridia, however, join a large bladder that is contractile and beats rhythmically for osmoregulation, since most rotifers are fresh-water animals.

Rotifers have a nervous system consisting of a brain lying dorsal to the pharynx, and several ventral nerve cords. There may be ganglia at the mastax and the foot, as well as connections to lateral antennae in some species.

Reproduction is highly specialized in rotifers. Only about 10 per cent of the population are males; in some species males are unknown. Females produce eggs that develop into embryos parthenogenet-ically. In other groups, the females produce two kinds of eggs: thin-shelled "summer" eggs and thick-shelled "winter" eggs that resist cold or drying. The thin-shelled eggs are haploid and may form females, but at times short-lived males are also formed and fertilize the females by hypodermic impregnation. These diploid eggs are thick-shelled and become females after overwintering.

Because of the small size of the rotifers, the nu-cleus divides a limited number of times at cleavage; a syncytium then forms for further growth. The nu-clear limits prevent regeneration and limit the num-ber of offspring. In many species the number of nu-clei are said to be constant within the species.

Ecology. There are about fifteen hundred species of Rotifera. Of these, about fifty are marine and pelagic, or epizoic on small crustaceans. Among the fresh-water forms, most are shallow-water or bot-tom-dwellers, but there are a number that parasitize other invertebrates. A few are terrestrial, but these are active only when water is present on plant leaves, and are otherwise dormant and encysted.

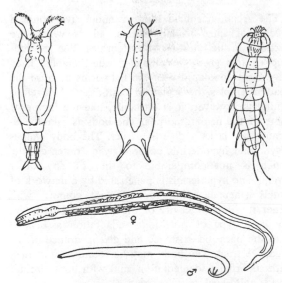

Fig. 118. Aschelminthes. *Top, Left to Right*–Rotifera, Gastrotricha, Kinorhyncha. *Bottom*–Nematoda

Gastrotricha. The gastrotrichs are a small group, related to the rotifers. Some are fresh-water; others are inhabitants of the marine intertidal regions. Gas-trotrichs are about the same size as rotifers, have tufts of cilia on the head region, and the cuticle cov-ering the body may be modified into scales, spines, or bristles. They glide about on ventral cilia much like flatworms. Protonephridia are usually present. The digestive tract is divided into a long pharynx

and a large stomach-intestine with a short rectum. Gastrotrichs are hermaphroditic, but in some species the males are degenerate and parthenogenesis is the functional means of reproducing. The eggs are re-tained in the female to develop. Dormant and non-dormant eggs are produced by some of the fresh-water species, as in the rotifers.

Kinorhyncha. The kinorhynchs are small (less than 1 millimeter), lacking in cilia, and with the body divided into head, neck, and trunk. The cuticle is divided into segmented plates that bear spines. The mouth is borne on an eversible cone armed with spines or bristles. Reproduction is dioecious and the males are not different from the females in appear-ance; gonads are fully developed in both males and females but fertilization and embryology have never been observed.

The kinorhynchs are marine, inhabiting algae and mud of the shallow-water littoral zone. Little work has been done on the group, although it is said that they are not difficult to find if one knows where to look.

Nematoda. The nematodes, or round worms, are among the most numerous animal groups in exist-ence. They are found inhabiting marine, fresh-water, and terrestrial environments, and are free-living or parasitic. The parasitic forms exhibit all degrees of dependence on the host, from spending only a minor part of the life cycle to complete dependence on hosts, with complex cycles. They include some of the most damaging of plant and animal parasites.

The nematodes are, almost uniformly, elongate slender worms of a size ranging from microscopic to several centimeters in length. The outer covering is a noncellular cuticle secreted by a syncytial epider-mis. Two pairs of longitudinal muscle bundles are present, but circular muscles are lacking; this pro-duces a whipping movement. The absence of true mesoderm precludes the development of muscle lay-ers in the body wall or around the gut. Ectomeso-dermal mesenchyme forms the muscle fibers pres-ent.

The mouth of nematodes is terminal and round, with cutting plates arranged in a triangle in some species. The anus is subterminal, with the gut di-vided into pharynx and a straight intestine. Pro-tonephrida are absent in nematodes; renette cells, which resemble them but lack the cilia of flame cells, empty into lateral canals that usually open to the outside in the middle body region. The nervous sys-tem consists of a brain ring around the esophagus, and longitudinal nerves. There are sensory bristles and papillae, as well as glandulo-sensory structures. Those near the mouth are called **amphids** and those near the anus are called **phasmids.** A few species have eyes.

Reproduction is dioecious; the females have paired tubular coiled ovaries which expand into uteri and join at the vagina. Some are oviparous, while others retain the eggs and are ovoviviparous. The males, which are usually much smaller than the females, have a single testicular tube and a short vas deferens leading to a seminal vesicle where the sperm are stored. The tube opens into the common cloacal chamber; accessory spicules are sometimes present to aid in copulation.

Nematodes are perhaps the most numerous plant parasites; a spoonful of soil from around the roots of garden plants will show masses of the worms under the microscope. Infections are also common in animals. The so-called hook worms are often found in the southern states, where children are apt to go barefooted, and in areas where feces disposal is poor. The eggs, or live larvae, are deposited with the feces and become rhabditiform free-living larvae. These may reproduce several times and then become long slender filaria worms, which penetrate the bare skin and migrate through the bloodstream to the lungs or gut where they settle.

In some species of intestinal worms the eggs may be ingested in contaminated food. *Necator* is typical of this latter type. Another means of infection is seen in the *Trichinella* worm, whose principal host is the pig. As is often the case when a parasite gets into the "wrong" host accidentally, the life cycle cannot be completed, but the larvae settle down in strategic locations which cause the new host much difficulty. When man eats poorly cooked pork, he ingests the *Trichinella* larvae which then become activated. They burrow through the gut or migrate up the gall bladder duct and encyst in any or all organs and in muscles. Eventually they become calcified as the host attempts to wall them off. Naturally such invasion of tissues causes severe strain on organ functions as well as great pain in the muscles. It is probable that the ancient Jewish prohibition of eating pork originated in the knowledge that severe symptoms were somehow associated with it. The taboos of many societies probably find their roots in similar circumstances; the wiser members found that making the act a religious prohibition was a better way to enforce their knowledge than education of their masses. It has been estimated that *Ascaris* infections in China deprive their human hosts of carbohydrates equivalent to 143,000 tons of rice a year.

Nematodes that do not inhabit the gut, but reside in the tissue of man, are for the most part transmitted by the bites of insects. The microfilaria offspring are sucked from the bloodstream and develop in the insect until they are ready to be injected or deposited on the skin of a new host by the fly. They migrate through the lymph channels and may settle in skin nodules, under the eyeball surface, or in the narrower lymph ducts, depending on the species. Since blocked lymph ducts cannot drain properly, the peripheral tissue becomes swollen and distended, producing such symptoms as elephantiasis. In some cases the only cure is to cut in and slowly pull the worm out, a difficult feat that encourages secondary bacterial infection. Control of the insects is the only effective method of eliminating tissue worms. In one species, open ulcers form through which the female worms give birth to live young when the host stands or washes in water, thus contaminating a stream.

Nematomorpha. These number only about eighty species, and are parasitic in insects. They are chiefly noted as the horsehair worms that were seen in horse watering troughs and were thought to appear by spontaneous generation. The larval stage occurs in insects; the adults are briefly free-living in fresh water, where they mate but are unable to feed.

PHYLUM ACANTHOCEPHALA

The Acanthocephala number about five hundred species of spiny-headed worms, all of which are parasitic in the intestine of vertebrates. The relationship of this group to others is questionable. They have been compared with the flatworms because the head resembles the scolex of cestodes and develops similarly; however, it is retractile, like the larvae of some of the nematomorphs. The body is pseudocoelomate but lacks a digestive tract. The body wall is a syncytial hypodermis covered by an outer cuticle, which is not comparable to that of any other group. The hypodermis is penetrated by a network of canals (lacunae), but they do not connect with either the interior cavity or the exterior. This is, no doubt, a type of nutritional adaptation associated with the parasitic existence and disappearance of the digestive tract. The body may be round or flattened, with superficial segmentation, and with both circular and longitudinal muscle fibers.

The sexes are separate, and fertilization is internal, with the male inserting a penis in the terminal gonopore of the female. The ovaries are simply egg masses collected into ovarian balls. Development is internal until the larval stage with hooks on the head is reached; then a shell is laid down and the "egg" extruded. The eggs pass out of the vertebrate host in the feces, and are then eaten by an arthropod intermediate host. They burrow through the intestinal wall and develop in the blood sinuses (hemocoels) until the arthropods are eaten by the vertebrate host.

PHYLUM ENTOPROCTA

The Entoprocta include about sixty species of pseudocoelomate animals that were formerly included in the Phylum Bryozoa. The remainder of the bryozoans are members of the Ectoprocta, which have been elevated to phylum status and separated from the entoprocts because the ectoprocts are coelomate. The relationships of the entoprocts to the other pseudocoelomate groups is obscure, although this phylum appears closest to the Rotifera.

Most entoprocts are sessile or attached, colonial, and marine; the colonial forms may be connected by stolons. One family is solitary and one genus is found in fresh water. All are quite small, from less than 1 millimeter to 5 millimeters in length.

The entoproct body consists of a stalk and a calyx, a bowl-shaped structure. The digestive tract of esophagus, stomach, intestine, and rectum lies in the calyx in a U, so that mouth and anus are close together on the face of the calyx. Solid, ciliated tentacles (mesenchyme-filled) extend from the edge of the calyx, encircling the mouth and anus. The tentacles can be rolled up on the face but cannot be pulled into the interior, as is the case in the ectoprocts. Calyx and stalk are covered by a thin cuticle. The calyx may be separated from the stalk by a partition so that the calyx can be shed and a new one grown.

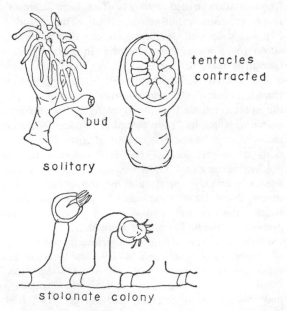

tentacles contracted

bud

solitary

stolonate colony

Fig. 119. Entoprocta

Budding is common in the entoprocts, producing colonies in some; the solitary forms also bear buds on the stalk that drop off to form new individuals. Some species are hermaphroditic while others are dioecious. The eggs are retained in the area of the anus, but how the sperm reach the egg is unknown. The free-swimming larvae produced by some species resemble the trochophore type seen in annelids. Protonephridia open by a nephridiopore near the anus, as does the gonopore.

Many entoprocts are epizoic, particularly on annelids or their tubes, on sponges, and on ectoproct colonies. Some have recently been found on crustacean gills by this author.

PHYLUM ANNELIDA

The remaining phyla to be discussed are all coelomate; either protostomes or deuterostomes. We assume a branching in evolution from the acoel body plan because it is evident that the groups departed early on separate lines of specialization. The annelids have the most generalized and regular coelomate body plan, elaborating on the basic wormlike ancestral structure. To this, annelids have added segmentation (or metamerism), and appendages.

Metamerism is more regular and advanced in the annelids than in the other phyla; the entire body is divided into segments partitioned internally by septa. The digestive tract, circulatory vessels, nerve cords, and excretory tubules must pass through the septa, but each segment is a fluid-filled compartment that can be expanded or contracted separately. The organs may also be segmentally arranged.

TABLE XIII

Classification of Phylum Annelida

Class Polychaeta	—with parapodia
Subclass Errantia	—free-swimming (*Nereis; Eunice*)
no orders, many families	
Subclass Sedentaria	—tube-living (*Sabella; Serpula*)
no orders, many families	
Class Oligochaeta	—with setae only (earthworms; *Lumbricus*)
Hirudinea	—(the leeches; *Hirudo*)

The annelids have a cuticle secreted by the epidermis. Beneath the epidermis is a layer of circular muscles and masses of longitudinal muscles. The coelom is lined with mesoderm, and the gut is surrounded by a muscle layer. In the Class Polychaeta fleshy paddle-shaped appendages (parapodia) occur, which are reinforced by stiff bristles. The parapodia serve partly for locomotion in swimming and are partly respiratory. They are lost in the terrestrial forms, the Class Oligochaeta. Paired bristles (setae or chaetae) provide movement for the latter group.

Polychaetes are considered to be the more primitive of the annelids. They are marine but commonly live in or on the muddy bottoms in quiet waters, so many people are entirely unaware of their existence. A large scoop of bottom mud may be screened to reveal a number of the worms, which range in size from 2 millimeters to more than a meter and may be brightly colored. The head may bear a variety of fleshy appendages called **palps** or tentacles; under a low-power microscope the "face" has an amusing appearance with the fleshy lips and teeth of an eversible prostomium (mouth), and the various appendages protruding. The free-swimming worms may be pelagic or simply crawl and burrow about the rocks and bottom fauna; some build burrows or tubes but leave them in order to breed or feed. The sedentary worms build tubes and burrows and remain there, feeding by pumping water through the tube or extruding mucus to trap food and pull it in. Many of the sedentary worms develop feathery tentacles about the head to aid in respiration, and are otherwise modified by their habitat.

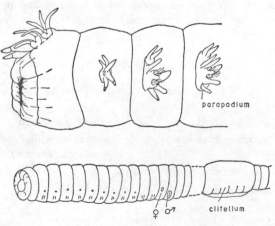

Fig. 120. Annelida. *Top*–Polychaete, anterior. *Bottom*–Oligochaete

The digestive canal of annelids typically consists of a pharynx, esophagus, stomach, intestine, and rectum, but in some species the different regions are not easily observed. Proteins, starches, and fats are apparently digested extracellularly, with absorption in the intestinal region. Muscular action of the intestinal wall aids in passing food in the coelomate animals.

The nervous system in annelids consists of a dorsal brain, connectives around the anterior gut, and a ventral nerve cord with segmentally arranged ganglia. Eyes are found on some of the Errantia. Chemoreceptor pits and statocysts also may be found.

The circulatory patterns, while varied, have a general plan. This consists of a dorsal vessel with anterior flow, and usually some contractile centers, or hearts. Connectives loop around the intestine and a ventral vessel flows to the posterior. Within each segment, in at least one polychaete studied, it has been found that blood flows from the dorsal vessel into the parapodia, down the parapodia to the ventral vessel, and back up around the intestine to the dorsal vessel.

Excretion may be accomplished by protonephridia with bulbous tips called **solenocytes** instead of the usual flame cells, or metanephridia may be present. The latter consist of a ciliated funnel or nephrostome, a long tubule, and a bladder. Either type of structure, or variations, may occur in different regions of the same animal, or in different species. The earthworms have pores opening to the exterior of each segment. Marine and aquatic worms secrete ammonia; terrestrial forms excrete urea.

Reproduction is dioecious in the polychaetes, and hermaphroditic, or monoecious, in the oligochaetes. Regeneration is well developed in the annelids; some marine worms regularly reproduce by a specialized fragmentation called **epitoky.** The hermaphroditic forms are usually not self-fertilizing. The reproductive tracts are well developed structurally, and copulation to exchange sperm occurs. In earthworms, a cocoon is secreted that is shed over the anterior end of the animal. This carries with it the eggs ejected from the fourteenth segment oviduct opening, plus the sperm, released at the opening in the fifteenth segment, which had been stored in the seminal receptacles at copulation. The cocoon is formed by secretions of the region of the body around the reproductive openings known as the clitellum. Cleavage is markedly spiral, and in the polychaetes a trochophore larva develops. In terrestrial oligochaetes the cleavage is modified by yolk, and development is direct, inside the cocoon.

Hirudinea. The leeches are primarily inhabitants of fresh water, although a few are marine and some are terrestrial. Again, those called terrestrial require much moisture in their environment, such as may be found in tropical rain forests. The body plan is remarkably constant, with the first two segments modified into an anterior sucker, and the last seven segments modified into a large posterior sucker. The body totals thirty-two segments, with smaller creases called **annuli** between segments. There are no parapodia and few have setae.

The coelom is greatly reduced by proliferation of mesenchyme; the circulatory system consists of a sinusoid hemocoel formed in the mesenchyme. Leeches are hermaphroditic and form cocoons as do the oligochaetes.

Leeches feed on dead animals and on small living invertebrates, but are chiefly known for the blood-sucking habits of certain species. They secrete a salivary enzyme that enables them to penetrate tissue and prevents coagulation. They may be ectoparasites on vertebrates. The digestive tract has a muscular pharynx to aid in sucking, and the gut is modified into many branches. The anus is dorsal because of the posterior sucker location.

ANNELID ALLIES

Three phyla, the Echiuroidea, Sipunculoidea, and Priapulida, are vermiform and have been variously assigned to the annelids, aschelminthes, and other affiliations at different times. They are not well known or numerous, and point up the fact that not all animals fit the major categories exactly.

Phylum Echiuroidea (or Echiurida). These are marine worms that burrow in sand or mud bottoms or in rock and coral crevices. The body is divided into an anterior proboscis and a saclike trunk. The proboscis, which contains the brain, is modified into a gutter shape and may be much longer than the animal's trunk or it may be very short. The animal feeds by spinning a mucus net from the proboscis as it backs into its burrow. It pumps water through the burrow, filtering detritus and plankton into the net, which it then detaches and swallows.

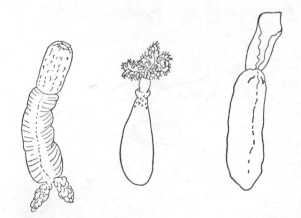

Fig. 121. *Left to Right*–Priapulid, Sipunculid, Echiuroid

The body wall is similar to that of the annelid, and setae may occur on the trunk below the mouth and around the anus. The digestive tract is coiled and has a strong gizzard and a terminal anus. Beside the anus are a pair of anal sacs of unknown function. Since there is a closed circulatory system in some species but not in others, these sacs may be for added respiratory surface, making contact with the hemoglobin present in cells in the coelom. The nervous system is also similar to that seen in annelids. Reproduction is dioecious, with external fertilization, and gametes are ejected through the metanephridial pores. The trochophore larva passes through a stage of segmented coelomic bundles before maturing into the nonsegmented adult.

Phylum Sipunculoidea. The sipunculids have a body divided into an anterior introvert and a tubular or saclike trunk. The brain is in the introvert region, which differs from the echiuroid proboscis in that it is retractable into the trunk and is equipped with tentacles, lobed or fringed, and spines or warts. Burrowers feed by ingesting sand and mud, as the terrestrial earthworms do, but nonburrowers trap food by the cilia and mucus on the tentacles. The anus is on the anterior trunk, since the digestive tract is coiled and U-shaped. There is one metanephridium, but there are unique cells called **urn cells** that are fixed or free and gather debris from the coelom. There is no circulatory system, although cells containing the respiratory pigment hemerythrin circulate in the coelomic fluid. Chemoreceptor sensory organs are found on the introvert surface.

Reproduction involves the formation of gametes on the coelomic wall and their extrusion through the nephridia. Fertilization is external and a trochophore larva forms, which does not show any traces of segmentation during subsequent development.

Phylum Priapulida. Until very recently the priapulids were classed with the pseudocoelomate animals, such as Aschelminthes, on the basis of early embryological studies. New studies with the electron microscope indicate that the priapulids are coelomate animals, and the cleavage is said to be radial. Radial cleavage would align this phylum with the deuterostomes, if there were not also a number of annelid-like features.

The body is a long sausage-shaped cylinder with an eversible armed mouth region, a segmented trunk, and caudal appendages that resemble bunches of grapes, of unknown function. The segmentation originates in the mesoderm, as in true coelomates, rather than being superficial as in pseudocoelomates. The epidermis is covered with a cuticle that is shed periodically; this is not characteristic of the deuterostomes. Protonephridia of the solenocyte type seen in some annelids occur in the priapulids. The nervous system is primitive, consisting of a nerve ring and a ventral cord. Gonads are associated with the nephridia, with connecting ducts. The embryos are said to resemble tiny rotifers. No circulatory system is present, but cells containing erythrin circulate in the coelom. These features overlap several phyla, and so the half dozen species must stand alone for the present.

PHYLUM MOLLUSCA

The mollusks are among the best known of all the invertebrate groups, probably because of the tremendous popularity of shell collecting. They are an old group, well differentiated in Cambrian times, and the fossil record is extensive because of the distinctive shell patterns. There are close to 100,000 species known, and they are widely distributed in nearly every habitat on earth. Some phyla have a few terrestrial species, but mollusks, like arthropods, have flourished on land, undergoing adaptive radiation.

Mollusks are unsegmented, coelomate organisms that apparently departed from the annelid stem line before the pronounced segmentation of the annelids developed. Segmentation is considered to be a device for aiding in locomotion in the worms; a muscular

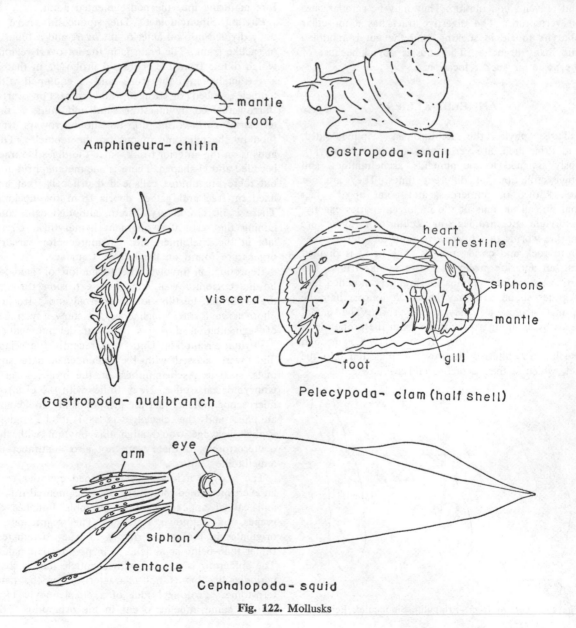

Fig. 122. Mollusks

foot and mucus glands accomplish this in mollusks. The muscular foot is topped by a dorsal visceral hump that is covered by a glandular tissue mantle. The mantle usually secretes a calcareous shell and aids in respiration. The head region is generally well developed and bears various sensory structures. The circulatory system is a combination of a muscular vessel or heart and coelomic sinuses (hemocoel).

TABLE XIV

Classification of Phylum Mollusca

Class Amphineura	—paired nerve cords, segmented plate shell (chitins)
Gastropoda	—coiled single shell (snails, limpets)
Pelecypoda	—bivalves (two shells) (mussels, oysters, clams)
Scaphopoda	—single tapering shell (tooth shells)
Cephalopoda	—shell coiled or absent (nautiloids, octopus, squid)

Since the classes are distinctly different, our discussion will deal with each class separately.

Amphineura. The chitins are slow-moving animals with reduced heads and eight moss-covered shell plates. They can be seen browsing on algae on the intertidal rocks. The heavy muscular foot clings by suction to the rocks, and a rasping rod in the mouth, called a **radula**, scrapes off the algae. This class was considered to be the most primitive of the living mollusks because the segmented plates and paired gills, nephridia, and gonopores might represent a more annelid-like ancestor. One related group, however, is wormlike but has no trace of segmentation.

In 1952, in the deep water off Costa Rica, several species of a more primitive form, previously known only as a Cambrian fossil, were dredged live. This species, named *Neopilina galatheae*, provided insight into the primitive features. It resembles somewhat a gastropod shell, with the foot and viscera of a chiton. Paired organs predate segmentation in animals; here there are five pairs of gills in the groove between mantle and foot, five pairs of retractor muscles, six pairs of nephridia, paired auricles in the heart, and two pairs of gonads. How similar this body plan is to the "ancestral" mollusk can only be surmised.

Gastropoda. The gastropods are mollusks that at some time in development undergo torsion, rotation of the viscera, so that the anus is anterior. This is thought to result from twisting in the larva, which enables it to withdraw its ciliated swimming organ; it may be due to the pressure of an evolutionary change in coiling, from a shell laid down like a garden hose to one that has the loops stacked. The subclass Prosobranchia of the gastropods shows the greatest degree of torsion, which causes a reduction in paired structures, eliminating those on the right side in some cases. All of the common marine and aquatic mollusks with a single peaked shell belong to this group, even though the coils may not be seen.

The members of the subclass Opisthobranchia have evolved away from the heavy shell and torsion characteristic, although torsion still occurs in the larvae. In many, the foot has been modified to enable the animal to swim and adopt a pelagic existence. "Sea butterflies" are of this group. Some retain the shell, others have a shell so reduced that it is covered by the mantle, and some have no shell at all. New "gills" have formed in two groups. In nudibranchs some have the dorsal surface covered with villi, called cerrata, for respiration, while others have a cluster of villi in one location.

Members of the subclass Pulmonata are those gastropods that have adapted to terrestrial habitats. Gills are absent and the mantle lining of the shell has become highly vascularized as a lung. *Helix,* the familiar garden snail, belongs to this group. All are hermaphroditic, with elaborate reproductive systems, but they are not self-fertilizing. Mating involves the launching of a calcareous dart from a sac into the tissue of the partner; this stimulates the male organ, and each transfers sperm in a spermatophore packet to the other by a penis. Gametes are produced by a combination ovotestis organ, but stored sperm disappear before ovulation occurs.

Pelecypoda. Members of this group are all bivalved, with the shells representing the right and left sides of the animal; the hinge is at the dorsal edge. The ventral foot gives the group its name, which means "hatchet foot." They are also known as Lamellobranchia because of their layered gills. A large clam can easily be examined for the principal anatomical features. A thin knife can be inserted between the halves, cutting the heavy muscles that hold it shut near the top of the shell. The body is covered by the soft mantle layer. This is fused near the dorsal posterior region into a ventral in-current siphon and a dorsal ex-current siphon. Water is circulated over the gill lamellae by cilia, which sort out debris and direct food toward the palps by the mouth. Water then exits along the dorsal edge of the gills. The gills are filaments that show varying degrees of fusion in the pelecypods. Clams have complete fusion. The posterior siphons allow pelecypods to burrow in sand or mud, head down, without blocking the water flow.

The foot is continuous with the soft visceral hump. The pharynx and stomach are embedded in a digestive gland, and the looped intestine may be covered by the gonad. The intestine passes through the pericardial cavity along the dorsal edge, and the rectum exits at the ex-current siphon. The circulatory system is fairly typical, consisting of two auricles and a ventricle that lies astride the rectum. The kidney is just above the gills.

Most pelecypods live in the intertidal or littoral waters, burrowing about in the sand or mud bottom, although some are dredged from deeper waters. Several species have become adapted to boring into wood or rock and concrete. The "shipworm," *Teredo*, is actually a pelecypod.

Scaphopoda. The tooth shells are burrowing marine mollusks with elongated shells open at both ends and the body correspondingly elongated. The head and sensory structures are reduced or absent; circulation is sinusoid. The digestive tract is similar to that of pelecypods, and water currents flow across the mantle in both directions, entering and exiting through the upper opening in the shell. The Indians used to employ these shells in decoration and for money.

Reproduction of pelecypods and scaphopods is similar; the animals are dioecious and fertilization is usually external, although some pelecypods retain the eggs to incubate on the gill chambers. A trocophore larva forms, which grows into a typical molluscan veliger larva. This form has a large velum which projects around and in front of the mouth and is ciliated for feeding and swimming. A miniature shell is present. In a few groups of pelecypods the veliger loses its velum and the shells hook on to the fins and gills of fish. There they develop as parasites, called glochidia, for some days and then drop to the muddy bottom to mature.

Cephalopoda. The cephalopods flourished early in geologic history, for fossils of some ten thousand species have been found. Presently, however, there are only about four hundred living species, although they include the largest invertebrates ever developed, and some of the fastest animals, rivaling the fish. There are three subclasses, the Ammonoidea, Nautiloidea, and Coleoidea.

The ammonoids are all extinct, and the only remaining nautiloid genus is the *Nautilus*. Squids and octopus represent the orders Decapoda and Octopoda respectively in the Coleoidea, the others of that subclass also being extinct. The ammonoids apparently had a primitive shell that was straight and cylindrical or cone-shaped. Coiling in the vertical plane gradually occurred, with some of the shells so loosely turned as not to touch between coils. The shell was added to at the anterior end, as the animal grew. Present-day nautiloids have a chambered shell that shows how partitions wall off the more distal portion as the animal grows. Many variations in shell form existed, of course.

The decapods include the squid genera *Loligo* and *Sepia*. In these, the shell is reduced to a horny pen beneath the mantle. The squid body consists of a foot that has become subdivided into eight short arms and two tentacles. The mantle has become a tough leathery covering with fins, and it is reinforced internally with a cartilage endoskeleton. This is the first endoskeleton seen in the invertebrates and consists of a brain case and wall plates. Behind the arms is a neck region bearing the beak or jaws and the eyes. The eyes are much like the vertebrate eye although they may not have been similarly derived. They may actually function more efficiently, since the human eye requires the muscles to change the shape of the crystalline lens to focus; the cephalopod eye muscles shift the entire lens back and forth. The members of the genus *Octopus* have a softer, rounded body and eight arms. They tend to be much more sedentary than the pelagic squid, lurking in crevices and around rocks to catch their prey.

In the squid the digestive system is well developed; it includes salivary glands, pharynx, stomach, digestive gland, and intestine, which empties anteriorly. Paired gills provide respiratory surfaces. The female has an ovary and a very large nidamental gland that secretes the egg covering. The oviduct has a flaring ostium. The male testis is small and has a small duct. At the tip of the duct the spermatophore packet with two bristles can sometimes be seen. One arm may be modified for transferring the packet to the female. The water currents come into the mantle cavity around the viscera and exit through a siphon in the central area. The mantle cavity in octopi is reduced but operates on the same principle. Both octopus and squid are able to swim backward rapidly by contracting the mantle cavity, which expels a jet of water for propulsion.

The circulatory system in cephalopods is closed and has capillaries. In addition to the systemic heart there are branchial (gill) hearts at the tops of the gills for added circulation.

The development of the eggs is direct, and is similar to that seen in some vertebrates. Cleavage is discoidal, with only a germinal disc cleaving over a large yolk, and then growing down to form a yolk sac.

Excretory organs consist of a pair of nephridia that open into the mantle cavity and are thought to excrete uric acid. The nervous system is highly developed, with a brain and several ganglia fused around the viscera, plus buccal and brachial ganglia. An ink sac for clouding the water opens off the rectum.

THE LOPHOPHORATE PHYLA

Three phyla, the Ectoprocta, the Phoronida, and the Brachiopoda, occupy an intermediate position between the annelid-arthropod allies and the echinoderm-chordate line. All three have an anterior prominence that bears the oral opening and is surrounded by ciliated hollow tentacles. The tentacles differ from those seen in more primitive organisms to the extent that the tentacle cavity is formed as an extension of the coelom. The bodies are divided into three regions—protosome, mesosome, and metasome—as is the case in the deuterostomes.

Ectoprocta. The Phylum Ectoprocta includes the coelomate members of the old Phylum Bryozoa. There are about four thousand living species, with many more fossil species recorded. Most are colonial and may form incrusting, stolonate, leafy, or arborescent colonies. There are two classes: the Phylactolaemata, which are fresh-water forms and have a horseshoe-shaped lophophore; and the Gymnolaemata, which are marine and have a circular lophophore. The arborescent forms are often mistaken for hydroids, which are found in the same communities. The incrusting forms cover rocks, shells, boat hulls, and other animals. The individual of a colony is called a zoid; each individual lies inside a calcareous or chitinous "box" that is lined with the coelomic tissue. The lophophore is retractable into the box (zooecium).

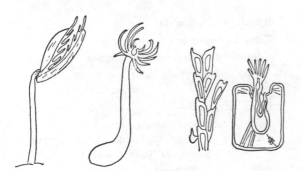

Fig. 123. Lophophorates. *Left to Right*–Brachiopod, Phoronid, Ectoproct (colony and zooecium)

Ectoprocta may be dioecious or hermaphroditic. Fertilization is usually internal, and development may be inside special structures called ovicells, which are zoids modified for the purpose. Cleavage is radial, but the larva is highly modified from the trochophore pattern. The larval gut degenerates and an entirely new gut forms, which obscures phyletic comparisons. There are no circulatory, respiratory, or excretory systems, no doubt because individuals are less than 1 millimeter long.

Phoronida. The phoronids are wormlike animals that build tubes in sand or on rocks of shallow marine waters. There are only about fifteen species, all bearing a lophophore that is horseshoe-shaped. The lophophore tentacles are modified into spiral folds that are typical of the group. As in other tube-dwelling animals, the digestive tract is U-shaped; the mouth is in the center of the lophophore, but the anus is outside the tentacles, as it is in ectoprocts. Phoronids are ciliary mucus feeders. The circulatory system is annelid-like and contains hemoglobin pigment. The nervous system consists of a nerve ring with branches to the tentacles, plus an epidermal nerve net.

The embryology of phoronids is open to question. Older literature indicates spiral cleavage and protostome development; the larva was interpreted as a modified trochophore. Newer research indicates radial cleavage, with traces of spiral pattern, and the larva is not accepted as a trochophore. This is another example of the intermediate position of these groups.

Brachiopoda. There are now only 250 living species out of some thirty thousand that flourished in earlier eras. Brachiopods have two shells, but the unequal valves represent the dorsal and ventral surfaces instead of the lateral surfaces, as in the pelecypod mollusks with which they were confused for many years. Most species are attached to the substrate of littoral waters by a stalk. Inside the shells the body bears a lophophore similar to that of the phoronids. The digestive system may be two-way, or it may have an anal opening. The circulatory system is open, with a muscular heart and no pigments.

The embryology is peculiar, with radial cleavage but protostome gastrulation. Mesoderm formation may be either enterocoelic or schizocoelic. The larvae are miniature adults, with mantle lobes for swimming, and the stalk is rolled up at the hinge area.

PHYLUM ARTHROPODA

The arthropods are bilaterally symmetrical, segmented coelomate animals with jointed appendages. There are more species and more individuals in the arthropods than in all other animals together.

Obviously this presents something of a problem in trying to provide a superficial examination of the group.

There is little doubt that the arthropods evolved from a segmental, vermiform ancestor; some of the more primitive forms are little changed from that condition. In the majority, reduction in length and segmentation has occurred, so that segmentation is never complete; it is seen in the segmental arrangement of nervous and circulatory structures and in breaks in the chitinous exoskeleton that has evolved from the cuticle.

The Onychophora are considered to show an intermediate condition between the annelids and the arthropods. Sometimes considered to be an arthropod class, they may also be placed in phylum status. *Peripatus* looks like a caterpillar, with annelid-like fleshy appendages which, however, are more ventral in position and bear claws, as the arthropods do. Respiration is through tracheal tubes, as in terrestrial arthropods. Whether these comparisons indicate the "missing link" status is an interesting question, depending on whether one applies the word "primitive" or "specialized" to the group.

The fossil record of the arthropods is well represented by some 39,000 species of trilobites, all now extinct. Both the major arthropod subphylum, the Mandibulata, which includes crustaceans and insects, and the Chelicerata, which includes spiders and scorpions, show similarities to the trilobites. Trilobites were ovoid, with a heavy exoskeleton. The body was divided into an anterior cephalon, an intermediate thorax divided into three longitudinal ridges, and a posterior pygidium. The appendages were alike, and had a walking branch and a respiratory branch (biramous). The cephalon was modified in various ways for plowing along the sea bottom. Trilobites disappeared when jawed fishes evolved; no doubt these fishes were able to feed on the previously invulnerable trilobites.

Class Crustacea. The divisions of the crustaceans show a series of progressive changes from the annelid type of structure in which similar appendages occur on every segment. The Branchiopoda, or "gill feet," are represented by the fairy shrimp, which have numerous appendages and lack any carapace, the exoskeletal shield. Others have developed a clamshell type of exoskeleton, which reduces the body length, and there are fewer pairs of trunk limbs. Copepods are longer and cylindrical but lack abdominal legs and have six pairs of thoracic limbs. The Cirripedia are the barnacles, which are attached and live in a calcareous cuplike structure. For years they were thought to be mollusks until the larva was observed metamorphosing. Here the animal in effect lies on its back and kicks through the opening.

TABLE XV

Classification of Phylum Arthropoda

Subphylum Trilobita
Subphylum Mandibulata
  Class Crustacea
    Subclass Branchiopoda
        Order Anostraca (fairy shrimp)
            Notostraca–shield carapace
            Diplostraca–2 valves
          Suborder Conchostraca (clam shrimp)
            Cladocera (water fleas)
      Subclass Ostracoda (seed shrimp, *Cypris*)
         Copepoda (*Cyclops*)
         Branchiura (fish lice, *Argulus*)
         Cirripedia (barnacles)
         Malacostraca (shrimp, crabs, lobsters)
      Superorder Peracarida–transparent thorax
        Order Mysidacea (possum shrimp)
           Isopoda (sow bugs)
           Amphipoda (sand hoppers)
      Superorder Eucarida–carapace over thorax
        Order Euphausiacea
           Decopoda–5 pairs of walking legs
         Suborder Natantia–swimmers (shrimps, prawns)
           Reptantia (lobsters, crayfish, crabs)

  Class Chilopoda (centipedes)
  Class Diplopoda (millipedes)
  Class Symphyla
  Class Pauropoda
  Class Insecta

Subphylum Chelicerata
  Class Merostomata
    Subclass Xiphosura (king crabs)
        Eurypterida (fossil giant water scorpions)
  Class Arachnida
    Order Scorpionida (scorpions)
        Pseudoscorpionida (false scorpions)
        Solpugida (sun spiders)
        Palpigrada
        Uropygi (whip scorpions)
        Amblypygi
        Araneae (spiders)
        Ricinulei
        Opiliones ("daddy long legs")
        Acarina (ticks, mites)
  Class Pycnogonida (sea "spiders")

The subclass Malacostraca includes the well-known shrimps, lobsters, and crabs. A typical body plan is seen in the common crayfish *Cambarus* that is found around fresh-water streams. The body is divided into two regions, a cephalothorax and an abdomen. The cephalothorax consists of a fusion of thirteen segments, each one bearing the following appendages. **Cephalic:** 1–antennule; 2–antenna; 3–mandible; 4, 5–maxillae; 6, 7, 8–maxillepeds. **Thoracic:** 9–chela (claws); 10, 11, 12, 13–walking legs. **Abdominal:** 14, 15, 16, 17, 18–swimmerets; 19–uropod. In the crabs, reduction of the appendages has been followed by extreme reduction of the abdomen.

The crustacean circulatory system is hemocoelic, with arteries but no veins; the respiratory pigment is hemocyanin. The excretory system consists of "green glands" that open at the base of the antennae. The digestive system has three regions: the foregut and hindgut have a chitinous lining and the central gut with a digestive gland is lined with endoderm.

The nervous system consists of a ventral nerve cord with segmental ganglia. The brain is a fusion of ganglia that encircle the esophagus, supplying the stalked eye, antennae, and statocyst. Reproduction is dioecious; in the crayfish there is a three-lobed ovary with oviducts leading to openings on the third walking legs. The male testis is also three-lobed and has two vasa deferentia that are highly coiled and terminate at openings on the fourth walking legs.

Crustaceans have adapted to a number of environments, although they are most numerous in shallow marine waters. The isopods have a few terrestrial species, and many of the crabs can remain away from water for some time. Some, like the coconut crab, are terrestrial, returning to water only to breed. The coelom is then highly vascularized.

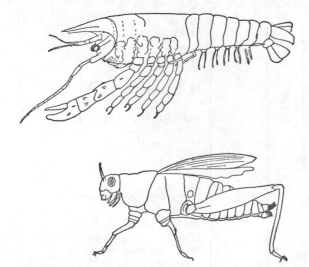

Fig. 124. Arthropods. *Top*–Crustacean. *Bottom*–Insect

Class Insecta. Table XVI gives some indication of the diversity of the insects. They represent the evolutionary peak of the arthropods, as do the cephalopods in the mollusks. Not only have they adapted to water and land but they have conquered the air. Insects show highly complex behavioral patterns, with many reflexes and some apparent associations. Although they are cold-blooded, insects have partly modified this by **fanning,** a period of waving the wings to warm the body before flight. The social life of honeybees and ants is well known for its strictly controlled role assignment to the various members of a colony.

The body plan of insects differs from that of the crustaceans in the separation of the head and thorax into distinct regions. The thorax is further marked into pro-, meso-, and metathorax, each bearing legs. The abdomen does not have appendages. The head is said to represent a fusion of six segments: the first bears two compound eyes plus ocelli (simple eyes); the second has sensory antennae; the third has the upper lip (labrum); the fourth has the mandibles; the fifth has maxillae or sensory palps; and the sixth has the second maxilla or the labium (lower lip).

The walking legs are the first two pairs; the third may be modified for jumping, as seen in the grasshopper, *Romalea.* The wings are not appendages but are outgrowths of the exoskeleton. Other external features include a tympanic membrane on the first abdominal segment, for auditory functions, and tracheal openings called **spiracles.** The tracheal tubules penetrate the body to allow air to circulate for oxygen exchange. Females have projections called **ovipositors** on the last segment for digging holes in which to lay eggs.

The heart is a single elongated vessel lying dorsally, and the body cavity functions as a hemocoel. No respiratory pigments are needed because of the tracheal system. The digestive tract begins with the mouth, which receives enzymes from large ventral salivary glands. A narrow esophagus leads to a thin-walled crop armed with spines, presumably to shred food. The crop is followed by a muscular gizzard, which ends the foregut. The stomach extends through the abdomen and is the major absorptive region, with paired hepatic caeca emptying enzymes into the lumen. The hindgut is short and lined with chitin. Excretory Malphigian tubules lie along the stomach, filtering the hemocoel and dumping the wastes into the midgut. Uric acid is excreted.

Myriapod Arthropods. Several classes of mandibulates are grouped in a nontaxonomic collection called myriapods. This group includes the Diplopoda, or thousand-leggers, and the Chilopoda, or centipedes. The slow-moving millipedes have two pairs

Table XVI

Orders of the Class Insecta

Metamorphosis	Orders and Common Names	Mouth Parts C-chewing S-sucking			Wings M-membranous R-roofed (resting over abdomen)	
					Fore	Hind
Ametabola: no metamorphosis	1. Diplura: japygids 2. Protura 3. Collembola: springtails 4. Thysanura: bristletails	C C C C	Apterygota: no wings			
Hemimetabola: young are nymphs, compound eyes, wings grow externally; metamorphosis gradual, incomplete.	5. Ephemeroptera: May-flies 6. Odonata: dragon flies	C C	Paleoptera: nonfolding wings	Pterygota: typically winged; some with wings reduced or absent.	M larger M nearly alike	smaller
	7. Isoptera: termites 8. Orthoptera: grasshoppers 9. Dermaptera: earwigs 10. Embioptera: embiids 11. Plecoptera: stone flies 12. Zoraptera 13. Corrodentia: book lice, psocids 14. Mallophaga: biting lice 15. Anoplura: sucking lice 16. Heteroptera: true bugs 17. Homoptera: aphids, scale insects 18. Thysanoptera: thrips	C C C C C C C C S S S S	Neoptera: modern, folding wings		Sexual forms alike, others wingless 4 or none leathery Hard, short Male winged, female wingless M, narrower None 4R, or none None None Half leathery 4, 2, or none Hairy fringe	thin thin, fanlike pleated, broader M
Holometabola: young are larvae, no compound eyes, wings grow internally; metamorphosis complex, complete.	19. Neuroptera: ant lions, snake, Dobson flies ✓20. Hymenoptera: ants, wasps, bees ✓21. Coleoptera: beetles, weevils 22. Mecoptera: scorpion flies 23. Trichoptera: caddis flies ✓24. Lepidoptera: moths, butterflies ✓25. Diptera: true flies 26. Siphonaptera: fleas 27. Strepsiptera: stylops	C C- S C C C S S S C			M, R, nearly alike M, 2 pairs or none Hard, veinless M, R, nearly alike M, R Wings with fine overlapping scales No hind wings None Male with hind wings, female none	M, folded

of walking legs on each segment. They have heavy jaws, and some have poison glands, although they are mostly herbivorous. When alarmed, the millipedes roll up (pill); this is also typical of such crustaceans as the isopod sow bugs.

The Chilopoda have one pair of long legs per segment and poison claws. They are swift-moving and predaceous, feeding on many kinds of small animals. They commonly live in and around ground litter and are sometimes seen in homes.

The Paurapoda are grublike forms with soft bodies and are closer to the diplopods than to other groups. There are twelve segments, nine of which bear legs. Exoskeletal tergal plates cover two segments each.

The Symphyla are more nearly related to the chilopods. They have twelve leg-bearing segments which are covered with 15 to 22 tergal plates. The last segment has spinnerets. They are very fast-moving, and feed on live or dead plants.

Subphylum Chelicerata. The chelicerates differ from the mandibulates in a number of ways. Usually there are two body sections, an unsegmented cephalothorax and an abdomen. There are no antennae and no true jaws. Instead, there are a single pair of chelicerae that may be designed to tear or to bite and inject poison, a single pair of pedipalps, and most have four pairs of walking legs. The abdomen usually has no legs. Respiration is by gills, lung books, or trachea. Excretion is by means of Malphigian tubules or by coxal glands at the base of the legs.

The Class Merostomata includes the giant horseshoe crabs, the *Xiphosura,* and the extinct giant water scorpions. The horseshoe or king crabs have a carapace reminiscent of the cephalon of the trilobites, and reach a length of 60 centimeters or more. They have five pairs of walking legs. The extinct Eurypterida had a similar cephalothorax, but the abdomen was subdivided and the middle portion bore six pairs of gills.

The eurypterids are believed to have been ancestral to the members of the Class Arachnida, named after the Greek mythological maiden Arachne, the spinner. This class includes the modern scorpions, the true spiders, and the ticks and mites. The arachnids show the evolutionary tendency toward fusion and reduction, with great modification of internal and external structures. At the extreme, the ticks have entirely fused the abdomen with the cephalothorax.

The Class Pycnogonida includes a small group of animals known as sea "spiders." Their relationships are questionable, however, and they are sometimes separated into phylum status. The body is long and narrow, with several segments. There are chelicerae, palps, a pair of short legs much modified to act as brood chambers for developing eggs, and four to six walking legs. Pycnogonids are marine; some crawl about on the hydroids and algae or bryozoa on which they feed, while others are able to swim with a flapping motion.

PHYLUM CHAETOGNATHA

The chaetognaths, or arrow worms, are significant because they represent the earliest offshoot of the deuterostome line that is firmly established in the echinoderm-chordate pattern. The organisms are small planktonic animals, predatory, and equipped with a spine-filled mouth. The adult body appears thin-walled and cuticular except for longitudinal muscles. Although the development is coelomate, the adult lacks a peritoneum and the cavity resembles a pseudocoel.

PHYLUM ECHINODERMATA

The echinoderms, or spiny-skinned animals, are the most numerous of the deuterostome invertebrates. Although the adults show radial symmetry, the larval forms are bilateral, indicating that radial symmetry is secondary. The cleavage patterns, gastrulation, and the dipleurula or other larvae of non-trochophore types relate the group to the chordate line of evolution. There is a dermal skeleton, a water vascular system, and gonads with associated ducts; kidneys are absent.

TABLE XVII

Classification of Phylum Echinodermata

Class		
Crinoidea	(sea lilies)	
Ophiuroidea	(serpent stars)	
Asteroidea	(starfishes)	
Echinoidea	(sea urchins and sand dollars)	
Holothuroidea	(sea cucumbers)	

The starfish is a familiar animal that lives in the littoral zone, creeping over rock or sandy bottom. It has a ring canal connected to lateral canals bearing tube feet. The tube feet have internal bulbs that can be contracted muscularly, causing the external tips of the feet to grip or detach from the surface.

Each of the five arms of the common starfish has a central groove through which the tube feet extend; the groove is flanked by a row of heavy spines that are extensions of the dermal plates beneath the skin. The stomach is in the center and can be everted through the mouth to engulf prey. The upper aboral surface is spiny also, and bears the opening sieve plate for the water vascular system. Respiration is by means of gill filaments that protrude through the skin and connect with the coelom. Gills are protected by pinching organs, tiny pedicellaria.

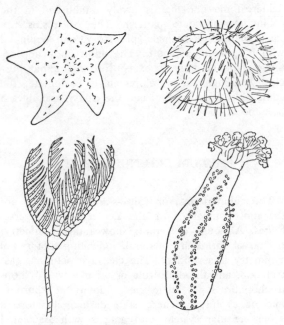

Fig. 125. Echinoderms. *Clockwise from Upper Left–* Asteroid, Echinoid, Holothuroid, Crinoid

The stalked crinoids are regarded by some as the most primitive echinoderms, and are represented in the fossil record. The serpent stars have a very small central disc and many long slender arms, some of which resemble annelids if only the arm is seen. The echinoids are the sea urchins, which have the same body plan as the starfish although the arms are fused into a common rounded test, bearing spines. The holothuroids are less like the other echinoderms, most of them having a soft body and an introvert with tentacles. The sea cucumber, which crawls slowly along the muddy bottoms feeding on plankton and detritus, has tube feet arranged in longitudinal rows along the baggy surface. There is a much-coiled digestive tract, and respiratory trees connected with the cloaca. The animal has the rather disgusting ability to eviscerate itself, disgorging intestine and

contents on the unwary who pick it up suddenly. The coelom and water vascular system contain coelomocytes, some of which contain hemoglobin, while others contain yellow or brown granules.

Echinoderms have great regenerative abilities; starfish can regenerate an entire body if the disc and one arm are present.

PHYLUM HEMICHORDATA

The hemichordates are marine worms that live in or on sand or mud. The body consists of a long proboscis in some, or a lophophore-type head with tube feet in others, plus a trunk region. For many years it was held that a stiffening rod in the proboscis was a notochord, but recent research denies this. There is a dorsal hollow nerve cord in some species, and gill or pharyngeal slits, both considered chordate characteristics. The larva is similar to the echinoderm type, which also helps to indicate that this group may have been on the chordate stem line or had a common ancestor.

Pogonophora. The pogonophores are the bearded worms that inhabit tubes at great depths. They are sometimes also given as an early offshoot of the same line as the hemichordates; the chief evidence for this lies in their distinctly enterocoelic mesoderm formation.

PHYLUM CHORDATA

The chordates include three subphyla: Urochordata, which includes the sessile tunicates; Cephalochordata, which includes the lancelets such as *Amphioxus;* and the Vertebrata.

The chordate characteristics include the presence of the notochord, a structure unique to the chordates, either in the adult or in the embryo. Paired gill slits or pharyngeal slits in the adult or embryo are also characteristic, as is the presence of a dorsal hollow nerve cord in larva or adult. The urochordate larva has all of these features, which are lost when the adult becomes sessile. Cephalochordates retain the notochord in the adult, as do the jawless fish in the vertebrates. Both groups also show a basic segmentation, which is considered to have arisen separately from the segmentation seen in the annelid-arthropod line.

The features of the various vertebrates have been discussed previously in the chapters on morphology and development.

The Eutheria are the major group of vertebrates today. They have conquered land as an environment, returned to the sea (whales), and taken to the air

TABLE XVIII

Classification of Subphylum Vertebrata

Class Cyclostomata	—cartilage skeleton (jawless fish; lampreys)
Chondrichthyes	—cartilage skeleton; jawed (sharks)
Osteichthyes	—bony endoskeleton; (modern bony fish, teleosts)
Amphibia	—cartilage and bone; lung breathers; moist skin (frogs; salamanders)
Reptilia	—cartilage; bone; lungs; horny skin (snakes; lizards)
Aves	—cartilage; bone; lungs; feathers; warm-blooded (birds)
Mammalia	—cartilage; bone; lungs; hair; mammary glands (man)

The Class Mammalia is divided into orders on the basis of reproductive patterns, dentition—the kinds and distribution of teeth, and appendages.

(bats). It remains to be seen whether their highly developed cerebral association centers can overcome their inheritance of the combative temperament, typical of the great apes.

Man is said to have evolved from a tiny shrewlike animal that existed 60 million years ago. Some 40 million years ago the separate lines leading to the apes, the monkeys, the lemurs, and the modern shrews appeared. Twenty million years ago separate lines were seen leading to the apes, Old World monkeys, New World monkeys (tail-swingers), tarsiers, lemurs, and shrews. Somewhat more than ten million years ago the ape *Proconsul* showed the animal where the great ape line diverged from the line leading to the manlike hominids. In the last few years research in Africa has revealed the existence of a creature called *Zinjanthropus*, whose bone fragments indicate that he may have been the first human. Dr. and Mrs. L. S. B. Leakey have spent years in Tanganyika searching for fossils in the area that they believe to be the cradle of man.

TABLE XIX

Classification of Class Mammalia

Subclass Prototheria	—egg-laying mammals (platypus, anteaters)
Subclass Theria	—placental mammals
Infraclass Pantotheria	—extinct; ancestral to remainder
Infraclass Metatheria	—marsupials with pouch (kangaroos)
Infraclass Eutheria	—true placental mammals
Order Insectivora	—shrews, moles, hedgehogs
Dermoptera	—flying lemurs
Chiroptera	—bats
Primates	—lemurs, monkeys, apes, man
Edentata	—sloths, anteaters
Pholidota	—scaly anteaters
Lagomorpha	—rabbits, hares
Rodentia	—rats, mice
Cetacea	—whales
Carnivora	—dogs, bears, cats
Tubulidentata	—aardvarks
Proboscidia	—elephants
Hyracoidea	—hyraxes
Sirenia	—sea cows
Perissodactyla	—odd-toed hoofs; horses, tapirs (ungulates)
Artiodactyla	—even-toed hoofs; pigs, cattle, deer (ungulates)

No doubt much more will be found in Africa if political conditions permit.

The next oldest humanoid was the fossil *Australopithecus*, found in Africa some years ago. The Java man, or *Pithecanthropus erectus*, was thought to have lived between 300,000 and 400,000 years ago in China and possibly North Africa. Peking man, *Sinanthropus pekingensis*, also found in China, is dated at about 500,000 years ago. Neanderthal man is shown as populating much of Europe during the last glaciation. He was not a handsome specimen, but his forehead was higher than his predecessors' and his brain case was large. He made stone weapons and had an extensive social organization. It was after this period that Cro-Magnon man, a subspecies of the modern *Homo sapiens,* began to arise and spread over Europe, between 20,000 and 50,000 years ago.

Before leaving the discussion of the animal groups, the author would like to mention the remarkable contributions to taxonomic and morphological zoology of Libbie Henrietta Hyman. She has published five monographic volumes, detailing the Invertebrata in the most comprehensive and modern treatment anywhere available. Her knowledge of zoology is probably unequaled in scope by any living authority, and her volumes form the basis of most of the present-day texts in the field. She has also written a comparative vertebrate text still in use some forty years after publication. Anyone wishing authoritative volumes on animal morphology would do well to seek her works.

Alfred S. Romer has written a most outstanding group of publications on the vertebrates and on vertebrate evolution. Again, these works are worthy of the attention of anyone interested in the evolution and diversity of the vertebrates.

CONCLUSION

Charles Darwin was born February 12, 1809, the same day as Abraham Lincoln, and each man had a tremendous impact in his field. Darwin's, of course, was biology. His early years were not spectacular, and his academic efforts at medicine and the clergy were unenthusiastic. In 1831, however, he was invited by one of his professors to become the naturalist on the explorative voyage of H.M.S. *Beagle,* which kept him away from England for five years.

The voyage offered Darwin unparalleled opportunity to observe and study at first hand the geology and natural history of many areas, and to make collections of specimens. He was a splendid observer, and prepared copious notes on all he saw. In 1837 he began to write the first of his journals, which culminated in the publication in 1859 of his *Origin of Species.* The first edition sold out almost overnight, stimulating a controversy that has never completely subsided.

Darwin concluded that evolution, or change, occurred in all living things by a process in which the best physically adapted specimens became the surviving, reproducing population. He saw that the organism had within it the potential, through selection, to adapt to whatever surroundings were available to him, if competition and physical conditions were not prohibitive. He felt that, by isolation of like species, continuing adaptation would eventually separate the species into different kinds, and he assembled masses of data that proved the truth of his conclusions.

We have discussed in some detail the genetic mechanisms that permit and encourage evolution, and we have compared anatomy, development, and physiology to show relationships among groups. Darwin's knowledge of geology led him to correlate the geologic record left by fossil remains throughout the world with the changing, evolving kinds of animals seen currently. In short, Darwin altered the concept of the biological world from one in which all different creatures were created at the same time, with similarities and differences purely accidental and of no significance, to one in which all creatures were related in some fashion. The differences between them arise from genetic changes and from the selective pressure exerted by their surroundings.

Ecology. Ecology is the study of living things in their relationship to their surroundings. This study is one that stems, directly or indirectly, from Darwin's works, and no biologist today examines specimens without considering the habitat and population relationships surrounding them. A considerable terminology has been developed to reflect the studies in the field, which Table XX partially covers. Biostatistics is a relatively new discipline that relates to ecology.

TABLE XX

Ecological Terms

MAJOR REGIONAL COMMUNITIES

Terrestrial:

Tundra—A northern terrestrial community with a cold Arctic climate where the ground is permanently frozen a few feet below the surface. There is no comparable land area in the South; the Antarctic is mostly ice.

Taiga—A northern terrestrial community immediately south of the tundra, characterized by a summer growing season sufficient to support the growth of conifers.

Alpine—Community on high mountains above the timberline with cold climate the year around.

Temperate—Communities with cold winters, warm summers, moderate precipitation supporting deciduous forests.

Rain forest—Terrestrial communities with abundant continuous rainfall and a prolonged growing season. Generally associated with the warm subtropical and tropical regions of the world.

Grasslands—Communities of the temperate and tropical regions where rainfall is insufficient to support forests but will permit heavy growth of grasses and herbs.

Desert—Communities with low rainfall, intense sunshine with hot days and cold nights.

Marine:

Littoral Zone—Area of shallow water covering the continental shelf to a depth of approximately 200 meters. May be divided into eulittoral zone and sublittoral zone.

Eulittoral Zone—Extends from high tide level to a depth of 40 to 50 meters.

Sublittoral Zone—From a depth of 50 to 200 meters.

Plankton—Microscopic-to-small marine plants and animals that float freely in the ocean, carried by currents and rising or falling with changes in water temperature or density.

Nekton—Free-swimming organisms of the oceans.

Benthos—Marine organisms fixed to or crawling on the sea bottom.

Neritic—Shallow water with light penetrations to the bottom (overlaps sublittoral).

Bathyal—Deep water (archibenthic) extending from sublittoral to between 800 and 1100 meters.

Abyssal—Deep waters beyond the bathyal zone, where solar light is lacking and the temperature does not rise above 5° C.

ZONES OF TEMPERATURE

Tropical—Summer temperature above 25° C.; winter, not much below 20° C.

Subtropical—No coral reefs present; temperatures near tropical.

Warm Temperate—Winter average for coldest month between 10° and 20° C.; summer maximum near 25° C.

Cold Temperate—Winter between 0° and 10° C.; summer mean about 20° C.

Frigid—0° C. in February, 10° C. in August, or the reverse.

It can be seen that altitude and depth may have the same biological results as latitude, longitude, or region, by providing the temperature range favorable to an animal and to plant growth on which the animal feeds. Observation and measurement have shown that certain plants and certain animals are just as indicative of zones as if signs were posted. The reverse, that in the zone certain biota will be seen, is also true. In this way the living plants and animals have made themselves at home on an inorganic and sometimes inhospitable earth.

GLOSSARY

GLOSSARY

Acoelomate—Without body cavity (coelom).

Adenosine Triphosphate (ATP)—Nitrogen base plus pentose sugar plus phosphate; acts as energy transport.

Aerobic—Oxygen-requiring form of respiration.

Allantois—Extra-embryonic membrane in reptiles, birds, and mammals; formed as an outpouching of the gut and invested with mesoderm. Acts as urinary bladder, or its vessels supply the placenta. Becomes adult bladder.

Allele (Allelomorph)—One of the alternative characteristics or genes that may occupy a given site on homologous chromosomes.

Alveolus (pl. alveoli)—Small sac or depression; the smallest unit of the lungs in mammals where oxygen exchange takes place.

Amino Acid—Organic acids that form the constituents of proteins.

Amnion—Extra-embryonic membrane present in reptiles, birds, and mammals formed of somatopleure folded around the embryo for protection; filled with amnionic fluid. An amniote is an animal possessing an amnion during embryological development.

Anaerobic—Respiration process without oxygen.

Anamniote—Vertebrates that lack the amnionic membrane during embryological development; chiefly fishes and amphibians.

Anaphase—Third stage of cell division in which chromosomes migrate toward the two poles.

Antibody—Substance produced by the body that reacts against a specific foreign protein introduced into the body.

Antigen—A foreign protein that causes the production of a specific antibody when introduced into the body.

Archenteron—The digestive cavity of the gastrula produced by invagination or ingression of endodermal cells.

Atom—Smallest unit of matter; composed of a nucleus containing neutrons and other particles of mass plus positively charged protons, and with negatively charged electrons circling the nucleus in orbits.

Auditory, Acoustic—Pertaining to the sense of hearing.

Autosome—Any chromosome other than the sex chromosomes.

Autotroph—Organism capable of manufacturing its own food by photosynthesis.

Axon—Nerve fiber extension of a nerve cell (neuron) that conducts impulses in a direction away from the cell body.

Bacterium (pl. bacteria)—Acellular organism, rod-shaped, round, or spiral; primitive, lacking a distinct nucleus with chromatin scattered through the cytoplasm.

Basement Membrane—Thin sheet that forms the underlying membrane of epithelial tissues.

Blastopore—Opening formed by invagination of the archenteron at gastrulation; may become the region of the mouth in some phyla, the anus in others.

Blastula—Hollow or solid ball of cells produced by cleavage prior to gastrulation.

Blepharoplast—Basal granule or activity center of a flagellum or cilium.

Bronchus (pl. bronchi); Bronchiole—Two tubes formed by division of the trachea to lead to each lung. Bronchioles are subdivisions of the bronchi.

Buffer—Substance capable of absorbing H ions (or OH), preventing large shifts in the *p*H of a solution.

Caecum—A blind pouch.

Calorie—A unit of heat; the amount of heat required to raise the temperature of 1 kilogram of water 1 degree C.

Carnivore—Meat-eating animal.

Carotene—Yellow or orange pigments that contribute to vitamin A formation.

Catalyst—Substance that accelerates a chemical reaction without becoming permanently altered or being part of the end product.

Cell—The basic structural and functional unit of living matter.

Centriole—One of two bodies that migrate to the cell poles and act as spindle poles during cell division.

Centrosome—A body, lying outside the nucleus, that contains the two centrioles during interphase.

Cercaria—Larval stage in flukes that encyst on the secondary intermediate host after formation in the small primary intermediate host.

Cerebellum—Region of the vertebrate brain responsible for muscular co-ordination.

Cerebrum—Center of association neurons in higher vertebrates; controls thought and memory.

Chloroplast—Small body in plants composed of layers of membrane and containing chlorophyll.

Choanocyte—Cells having protoplasmic collar and flagellum; found in sponges.

Chondrocyte—A cartilage cell.

Chorion—Outermost extra-embryonic membrane in reptiles, birds, and mammals; composed of ectoderm and mesoderm; remains in contact with uterine wall in placental animals.

Chromatid—A single-strand chromosome.

Chromatin—Stainable material of which chromosomes are composed.

Chromonema (Chromonemata)—Chromatid stretched out in threadlike shape.

Chromoplast—A saclike vesicle containing layers of membrane and colored pigments that aid in food manufacture.

Chromosome—Body in the nucleus that contains the genetic code; stainable during mitosis or meiosis as filaments or rods. The number and kind are constant for a given species.

Cilium (pl. cilia)—Short filament organelle on the surface of many cells; locomotory organelle in some protists and circulatory organelle on fixed tissues. Short flagellum.

Cleavage—Repeated cell division of the egg after fertilization until a ball of cells—blastula—is formed.

Cochlea—Part of inner ear, coiled like a snail shell. Houses auditory receptors that translate vibrations into nerve impulses.

Coelom—Body cavity or space between outer body wall and gut; true coelom is lined with mesoderm germ layer.

Coenzyme—Factor required to activate an enzyme.

Colon—Large intestine preceding the rectum and anus.

Colonialism—The state of individuals who remain joined together structurally.

Commensal—Literally "eating at the same table"; organism living with a host organism that feeds at the same time but does not benefit or harm the host.

Cyclosis—Circular streaming of the cytoplasm inside a cell.

Cytochrome—Hemoprotein that acts as electron transport in cellular oxidation within the mitochondria.

Cytology—The study of cell structure.

Cytoplasm—Cell protoplasm exclusive of the nucleus and organelles.

Dendrites—Filamentous outgrowths of neurons that conduct impulses from the periphery to the cell body.

Deoxyribonucleic Acid (DNA)—Molecules composed of ribose sugar that lacks one oxygen, linked to nitrogen bases of purine or pyrimidine, and connected by phosphate links. Believed to carry the genetic code by means of repeating sequences of molecules.

Desmosome—Rounded projections of the cell membrane that aid in interlocking the adjacent cells.

Diabetes—A disease, marked by deficiency of insulin, that prevents splitting of bonds of sugars for digestion.

Diffusion—Equalization of particles to form uniform concentrations; results from random movements of molecules.

Dioecious—Having separate individuals for male and female sexes; not hermaphroditic.

Diploidy—A double set of chromosomes; twice the complement of the gametes.

Ductus Arteriosis—Connection between the aortic (fourth) arch and the pulmonary (sixth) arch; open in fetus, normally closes at birth.

Duodenum—Area of small intestine at outlet of the stomach; bile duct and pancreatic duct empty into it.

Ectoderm—The germ layer that produces the epidermis and nervous system.

Embryo—Early developmental stage of an organism following cleavage and gastrulation.

Endoderm (Entoderm)—Germ layer that produces the lining of the digestive tract, liver, lungs, and bladder.

Endoplasmic Reticulum—System of transport tubules formed as an extension of the cell membrane into the cytoplasm.

Endosome—Structure in some primitive cells that contains the chromatin.

Enteric—Pertaining to the intestine.

Enterocoel—Coelom formation by means of outpouching of the gut endoderm.

Enzyme—A protein, produced by an organism, that acts as a catalyst.

Epidermis—The outermost layer of tissue of an organism.

Epididymis—A much-coiled tube that collects sperm formed in the testes.

Epithelial Tissue—Sheetlike tissue that forms the covering or lining of organs.

Esophagus—Specialized area of the digestive tract between the pharynx and stomach.

Estrogen—Female sex hormone produced by the primary follicle.

Estrus—Period of sexual excitation at the time of ovulation to ensure mating and fertilization of the egg. In some animals it is accompanied by a flow of mucus.

Eustachian—Tube connecting the throat with the middle ear and derived from a pharyngeal cleft.

Fetus—Stage of growth and development following embryonic stage and preceding birth.

Fibril—A small strand or filament produced and remaining inside a cell.

Fibrin—Coagulated blood protein that forms the mesh for blood clots.

Fibroblast—Connective tissue formative cell.

Fibrocyte—Basic mature connective tissue cell.

Flagellum (pl. flagella)—Whiplike organelle of protists and male gametes for locomotion. Composed of a circle of nine double rods and center of two rods.

Flavin-Adenine Dinucleotide (FAD)—Hydrogen acceptor in glycolysis.

Gamete—Haploid sex cell; sperm or ovum.

Ganglion—Cluster of neuron cell bodies outside the central nervous system.

Gastric—Pertaining to the stomach.

Gastrula—Developmental stage after blastula has invaginated to form archenteron; or primitive gut stage at which germ layers become fixed.

Gene—Segment of a chromosome that governs production of one enzyme in the genetic code.

Genotype—The genetic factors or sets of genes that determine inheritable patterns.

Gestation—Period of offspring development in the uterus.

Glial Cells—Several types of cells that act as supportive cells for the neurons.

Glycogen—Storage product formed by dehydration and linkage of saccharides.

Golgi Bodies (Golgi Apparatus)—Intracellular vesicles of a secretory nature; thought to be formed by the endoplasmic reticulum.

Gonad—Reproductive organ of either sex.

Ground Substance—Material produced by the cells and deposited between cells instead of intracellularly.

Haploid—A single set of chromosomes; the number found in the mature gametes.

Hemoglobin—Red pigment with oxygen-carrying ability combined with protein; found in red blood corpuscles.

Hepatic—Pertaining to the liver.

Heterotrophs—Animals that depend upon other organisms to produce food.

Histology—The study of tissues.

Histones—Unique proteins found on the chromosomes; thought to control production of the enzymes by the genes.

Hormone—Substance produced by a given tissue or organ that is transported by the blood to another organ to stimulate or inhibit a function.

Hydrocarbons—Organic compounds formed from carbon and hydrogen.

Hypertonic—Possessing a greater concentration of particles or molecules than the surrounding fluid separated by a semipermeable membrane barrier.

Hypophysis (Pituitary)—Ventral projection from the diencephalon of the brain that produces several hormones.

Hypotonic—Possessing a lesser concentration of particles or molecules than the surrounding fluid when separated by a semipermeable membrane barrier.

Ingestion—The intake of food.

Ingression—In-wandering of cells; one method of gastrulation.

Insulin—Hormone, produced by the Islet of Langerhans cells in the pancreas, that enables animals to split the glucose molecule.

Invagination—The pushing inward of a layer of tissue to produce a sac or hollow.

Invertebrates—Animals without backbones.

Ion—An atom electrically charged because of a deficiency or excess of electrons.

Isotonic—Equal concentration of particles or molecules on either side of an osmotic membrane.

Isotope—One of several forms of an element differing in nuclear weight but not in activity.

Kidney—Organ of osmoregulation and excretion in vertebrates.

Kinetochore (Centromere)—Activity center of a chromosome; the area where spindle fibers attach during cell division.

Lacuna—Space within cartilage or bone matrix in which the cell body resides.

Lamellae—Layers; concentric rings deposited in bone formation. Also, gill layers in clams.

Larva—Stage of development in animals to enable them to be self-feeding before reaching the typical adult anatomy or form.

Leucocytes—White blood cells that function to surround foreign bodies and bacteria that invade the body.

Lipid—Fat or pertaining to fat.

Lymph—Body fluid in the tissues outside the blood vessels. It is returned to blood vessels by ducts and escapes from capillaries.

Lymphocyte—White blood cell thought to function in the production of antibodies.

Lysosome—Cell organelle capable of secreting hydrolytic enzymes that break down organic materials.

Macrophage—Cell capable of engulfing or ingesting particulate matter.

Mast Cells—Cells, found in connective tissue, that are "stuffed" with secretory granules.

Maxilla—Part of the upper jaw in vertebrates; the head appendage in arthropods.

Medulla—Region connecting the brain to the spinal cord. Also, the interior tissue of an organ.

Meiosis—Process of cell division that produces the haploid chromosome number; seen only in gametogenesis in the gonads.

Menstruation—The process in the primate uterus that produces a sloughing off of the uterine mucosa with blood and fluid pooled beneath; occurs every 28 days if the egg is not fertilized during the ovarian cycle.

Metabolism—The processes in the body that produce energy and build tissue.

Metaphase—Second stage of cell division in which the chromosomes are aligned in the equatorial plane of the cell.

Microsome—The site of protein synthesis in the cytoplasm. Also, the lightest weight fraction of cells that have been homogenized and centrifuged; includes endoplasmic reticulum and ribosomes. "Microsome" is sometimes used to mean "ribosome."

Microvilli—Minute finger-like projections seen in the cell membrane.

Mitochondrion (pl. mitochondria)—Cell organelle that is the site of cellular respiration with the release of energy.

Mitosis—Process of cell division that produces two identical daughter cells with the diploid chromosome number.

Morphogenesis—The development of adult shape and form.

Mucosa—Tissue layer that secretes mucus.

Mucus—Sticky substances secreted to protect membranes; a mucopolysaccharide (sugar plus amino acid).

Mutualism—Symbiosis in which both animal and host are benefited by the association.

Myelin—Fatty material secreted by neurolemmal cells to form sheathing around axons of neurons in the central nervous system.

Myofibrils—Contractile fibrils within cells, especially muscle cells.

Nephridium—Osmoregulatory or excretory unit of invertebrates.

Nephros—Functional unit of excretion; kidney.

Neuron—Nerve cell that transmits electrochemical nerve impulses.

Nicotinamide-Adenine Dinucleotide (NAD)—Hydrogen acceptor in glycolysis.

Node—A swelling or concentration of cells, as in lymph node. Also, a joint.

Nuclear Membrane—Membrane enclosing the nucleus; may be an extension of the cell membrane.

Nucleic Acid—Large molecules composed of linked nucleotides; DNA and RNA.

Nucleolus—Body in the nucleus that contains RNA; always a function of a specific chromosome.

Nucleoside—A molecule composed of a 5-carbon sugar plus a purine or pyrimidine.

Nucleotide—A nucleoside plus phosphate: pentose sugar, plus purine or pyrimidine, plus phosphate.

Nucleus (atom)—The central mass of an atom around which the electrons orbit.

Nucleus (cell)—A body, present in most cells, that contains the chromosomal material, nucleolus, and nucleoplasm. It is surrounded by the nuclear membrane.

Olfactory—Pertaining to the sense of smell.

Oöcyte—Female germ cell that is undergoing meiotic division to form gametes.

Optic—Pertaining to the sense of vision.

Organ—A group of tissues joined structurally to perform a given function or functions.

Organelle—Structurally specialized area of a single cell, adapted to perform a given function.

Osmosis—The movement of water through a semipermeable membrane in the direction of the greater concentration of molecules.

Osteocyte—A bone cell.

Ovary—Female reproductive organ which produces eggs.

Oviparous—Reproduction in which development of the fertilized egg occurs outside the mother's body.

Ovoviviparous—Reproduction in which fertilization and development of the egg occur inside the body of the mother, but without any connection to the maternal body for nutrition.

Ovulation—The release of a developed ovum by the ovary.

Oxidation—The chemical process of adding oxygen to, or removing hydrogen from, a compound.

Parasite—Organism that lives in or on another animal to the detriment of the host.

Parasympathetic—Portion of the autonomic nervous system, mainly composed of extensions of the tenth cranial nerve and the caudal spinal nerves. Neuromuscular junction activated by acetylcholine; suppresses heart, stimulates digestion.

Parathyroid—Glands, originating in the pharyngeal pouches, that control calcium metabolism.

Parenchyma—The cells of an organ that perform the principal function of that organ.

Parthenogenesis—The development of the egg without fertilization.

Pathogenic—Disease-producing (organisms).

Perichondrium—Fibrous connective sheath around a cartilage structure.

Periosteum—Fibrous connective sheath around bone.

pH—Coefficient expressing the hydrogen ion concentration; a measure of relative acidity or alkalinity.

Phagocytosis—The ingestion of particulate matter by cells.

Pharynx—Portion of the digestive tract connecting the mouth (buccal cavity) with the esophagus.

Phenotype—An organism's physical appearance that results from its genetic makeup.

Physiology—The study of function and living processes.

Pinocytosis—The taking in of fluids by cells by means of surrounding the fluid droplet.

Placenta—Complex of tissues, formed by contact of extra-embryonic membranes with uterine mucosa, to nourish the embryo.

Plasma Cells—Cells in connective tissue that are thought to produce antibodies.

Platelet—Blood cell fragments suspended in the blood plasma.

Polyp—One form or body type of coelenterate consisting of an attached or sessile sac or stalk with tentacles surrounding the mouth. Corals and anemones are typical.

Progesterone—Hormone produced by the corpus luteum on the ovary after ovulation; prevents menstruation if fertilization occurs.

Prophase—First stage of cell division in which chromosomes become stainable and the spindle forms.

Pseudocoele—Body cavity between body wall and gut, but not lined with germ layer mesoderm; remnant of the blastocoele.

Pseudopodia—"False feet"; temporary projections of protoplasm in cells to produce amoeboid motion.

Pulmonary—Pertaining to the lungs.

Redia (pl. rediae)—Larval stage of fluke life cycle.

Renal—Pertaining to the kidney.

Respiration—Intracellular release of energy for metabolism in the mitochondria; may be with or without oxygen. Also refers to "breathing."

Reticulum—A network or meshwork.

Retina—Sensory layer of the interior of the eyeball.

Ribonucleic Acid (RNA)—Molecules that transport genetic information and act as templates for protein synthesis.

Ribosomes—Submicroscopic particles that are the sites of protein synthesis (sometimes called microsomes).

Saccharide—A sugar.

Sarcoplasm—The cytoplasm of muscle cells.

Schizocoel—Formation of coelom by splitting of the mesoderm germ layer.

Scrotum—Pouches in the skin in the groin area into which the testes descend, in mammals.

Septum (pl. septa)—Membrane or partition separating regions of a cell or structure.

Sinus—A cavity or hollow for fluid collection.

Somatic—Pertaining to the body.

Somatopleure—Body layer composed of ectoderm invested with mesoderm.

Somite—Series of tissue bundles that form body musculature.

Spermatocyte—Male germ cell in the process of developing into a gamete by meiosis.

Splanchnopleure—Body layer composed of endoderm invested with mesoderm.

Stroma—The connective tissue that provides support and bulk for the functional cells (parenchyma) of an organ.

Substrate—Reactants that are joined by an enzyme.

Substratum—Layer underlying another, the base.

Symbiosis—State in which one animal lives with a host animal; the term does not imply benefit or harm to either.

Sympathetic—Portion of autonomic system composed of the spinal nerves; neuromuscular junction stimulated by epinephrine (=adrenin, adrenaline). Stimulates heart, slows digestion.

Synapse—Microscopic gap between axon and dendrite of adjacent neurons; glial cells may secrete acetylcholine hormone that transfers impulse across the gap.

Syncytium—Multicellular tissue in which the cell nuclei remain but the cell boundaries disappear.

Taxonomy—The classification of organisms based on natural relationships.

Telophase—The last stage of cell division in which new nuclei and the cell membranes form.

Testis (pl. testes)—The male reproductive organ that produces sperm.

Thorax—Portion of the animal body between head or neck and abdomen.

Thrombocyte—A blood platelet that aids in clotting of blood.

Thymus—A gland, located in the throat and formed by the pharyngeal pouches, that initiates antibody production by forming lymphoblast cells. Large in infants but decreased in adults.

Thyroid—A gland that forms as a ventral pocket of pharynx and migrates to the throat; produces the hormone thyroxin which stimulates metabolism.

Tissue—A group of similar cells that are linked together to perform a given function.

Trachea—Main "windpipe" leading from pharynx to the lungs.

Umbilicus—Body stalk that connects abdomen of embryo to the placenta, which is in contact with the uterine wall of the mother; contains arteries and veins.

Undifferentiated Cells—Cells that have not yet become permanently committed as to fate or function; may differentiate into one of a number of kinds of cells.

Ureter—Duct carrying urine from the mammalian kidney to the bladder.

Uterus—Chamber for incubation of developing embryo, formed by enlargement and fusion of ducts leading from the ovaries.

Vacuole—Saclike vesicle within cells containing liquid or solid matter or both; may be permanent or transitory.

Vagus—The tenth cranial nerve, originating in the medulla and supplying the viscera as the major portion of the parasympathetic system.

Vertebrates—Animals with backbones.

Viviparous—Reproductive development in which the ovum becomes attached to the maternal uterus and receives nutrition from it during development and growth.

Zygote—A fertilized egg.

Zymogen Granules—Secretory granules sometimes observed in Golgi bodies.

INDEX

INDEX

A

Acanthocephala, Phylum, 162
Aerobic glycolysis, 109–11
Agglutination, 90
Alcohols, definition, 25
Alleles, 76, 81
Allen's Rule, 95
Amino acids, formation of, 27
Amphineura, Class, 167
Anaerobic glycolysis, 108–9
Anaphase:
  in meiosis, 78
  in mitosis, 65
Anhydro bond, definition, 28
Animal kingdom, survey, 145–76
Animal kingdom, table of, 148–49
Annelida, Phylum, 163–65
Antibody, 90
Antigen, 90
Aortic arches, 139–40
Aristotle, 13, 16
Arrhenius, Svante, 21
Arthropoda, Phylum, 169–73
  table of classification, 170
Aschelminthes, Phylum, 160–62
Atomic number, definition, 22
Atomic orbits, 22
Atomic weight, definition, 22
Atoms, definition, 21
Autonomic nervous system, 136
Autotrophs, definition, 34
Axons, 61, 131

B

Back cross, 80, 87
Belon, Pierre, 17
Bermann's Rule, 95
Binomial system of nomenclature, 13
Bisexuality, 73–75
Blood, 58–59
Blood types, 90
Bone, 58
Brachiopoda, Phylum, 169
Budding, as reproductive method, 74, 111

C

Carbohydrates, 26

Carboxyl group, definition, 26
Carotid bodies, 138
Cartilage, 57
Catalyst, definition, 28
Cell, 41–62
  definition, 24
  food for, 43
  functions of, 46–50
  membrane, 43
  theory, 41–42
  ultrastructure of, 43
Centrioles, 45
Centrosome, 42, 45
Cephalopoda, Class, 168
Cerebellum, 130, 133
Cestoda, Class, 158
Characteristics of animal kingdom, 145–47
Chelicerata, Subphylum, 173
Chemosynthesis, 34
Chlorophyll, 34
Chloroplast, 34
Chondrocyte, 57
Chordata, Phylum, 174–76
Chromatin, 64
Chromosomes, 42
  nature of, 66–73
Circulation, in embryo, 142
Circulatory system, 139–43
Classes, in nomenclature, 14
Cleavage, in reproduction, 118–19, 123–24
Cnidaria. See Coelenterata
Coacervate, definition, 32
Coelenterata, Phylum, 155–56
Collagenous fiber, 55
Colloids, 32
Colonialism, in cells, 47
Columnar epithelium, 51
Compound, definition, 22
Connective tissues, 54–60
Cranial nerves, 134–35
Crustacea, Class, 170–71
Ctenophora, Phylum, 156–57
Cuboidal epithelium, 51
Cytoplasm, 45
  definition, 64

D

Darwin, Charles, 79, 82, 176
Deme, 92

Democritus, 19
Dendrites, 61
Dense connective tissues, 56–57
Deoxyribonucleic acid. See DNA
Determination, in organic development, 126–28
Developmental anatomy, 129–43
Diakinesis, stage of meiosis, 77
Diencephalon, 133
Diffusion, 99–100
Digestion, 105–11
Digestive glands, 139
Digestive system, 136–39
Diploidy, 74
Diplotene, stage of meiosis, 76
DNA, 30, 68–73, 126
Dominance, in heredity, 79–82

E

Earth, theories of origin of, 20
Echinodermata, Phylum, 173–74
Echiuroidea, Phylum, 165
Ecology, 176
Ectoprocta, Phylum, 169
Elastic cartilage, 57
Elastic fiber, 55
Electromagnetic spectrum, 44
Electron microscope, 43
Electrons, definition, 22
Element, definition, 21
Embden-Meyerhof pathway, 108
Endocrine glands, 52–54
Endocrinology, 54
Energy, 24
Entoprocta, Phylum, 163
Enzyme inhibitors, 73
Enzymes, definition, 28
Epithelial tissues, 50–54
Erasistratus, 16
Esophagus, 139
Estrus cycle, 116
Eugenics, 83, 95–96
Eustachi, Bartolomeo, 16
Exocrine glands, 52–54

F

Families, in nomenclature, 14
Fat cells, 54
Fats, 26

Female reproductive organs, 114–15
Fertilization, 76
  in reproduction, 117
Fibroblast, 54
Fibrous cartilage, 57
Fission, as reproductive method, 111
Flemming, Walther, 63

G

Galen, 16, 17
Gametes, 74
Gastropoda, Class, 167
Gastrotricha, Class, 161
Gastrulation, 119–24
Gel state, 32
Gene pool, 93
Genes, 42, 80
  nature of, 66–73
Genetic drift, 93
Genotype, 81
Genus, in nomenclature, 13
Germ cells, 75
Glandular epithelium, 52
Gloger's Rule, 95
Glossary, 181–86
Golgi apparatus, 46
Gonads, 78, 113
Ground substance, 54, 55

H

Haploidy, 74
Hardy-Weinberg Law, 93
Harvey, William, 17
Haversian systems, 58
Heart, 141
Hemichordata, Phylum, 174
Heredity, 79–96
Herophilus, 16
Heterotrophs, 35
Heterozygous, 80
Hippocrates, 15
Histology, 49–50
Histones, 72
Hormones, 54
Hoyle, Fred, 20
Hyaline cartilage, 57
Hybrids, 80
Hydrocarbons, definition, 25
Hydrogen ion concentration, 97
Hydroxyl group, definition, 25

I

Independent assortment, 81
  law of, 83
Inheritance, polygenic, 88
Inorganic compounds, 26
Insecta, Class, 171
  table of orders of, 172

Intestine, 139
  large, 106
  small, 105
Invertebrates, 17
Ion, definition, 22
Ionic bond, 22
Isotopes, definition, 22

K

Keratinization, 52
Kinorhyncha, Class, 161
Krebs Cycle, 110–11

L

Lacuna, 57
Langerhans, islets of, 139
Leeuwenhoek, Anton von, 20, 41
Leptotene, stage of meiosis, 76
Life, theories of origin of, 21
Linnaeus, Carolus, 13
Loose connective tissues, 55–56
Lophophorate phyla, 169
Lung circulation, 141
Lungs, 138–39
Lymphatic system, 143
Lysosomes, 46

M

Macrophages, 54
Male reproductive system, 115
Malthus, Thomas, 96
Mammalia, table of classification of, 175
Mast cells, 55
Matter, properties of, 31–32
Medulla, 130, 133
Meiosis, 75–78
Membranes, protective, 124–25
Mendel, Gregor, 79ff.
Mental deficiency, 96
Mesencephalon, 129
Mesenchymal cells, 54
Mesozoa, Phylum, 153
Messenger RNA, 69–70
Metabolism, 102–5
Metaphase:
  in meiosis, 78
  in mitosis, 65
Microsomes, 46
Microtome, 49
Microvilli, 46
Midbrain, 133
Mitochondria, 45–46
Mitosis:
  in animals, 38, 64–66
  in plants, 66
Molecule:
  definition, 22
  properties of, 23

Mollusca, Phylum, 166–69
Monera, 37–38
Morgan, Thomas Hunt, 42
Morphology, definition, 13
Mucigen, 52
Muscle, 59–60
Mutations, 73, 82, 89
Mutualism, in cells, 47

N

Natural selection, 93
Nematoda, Class, 161
Nematomorpha, Class, 162
Nemertina, Phylum, 159–60
Nervous system, 129–36
  cells of, 60–61
Neuron, 130
Neutrons, definition, 22
Nitrogen cycle, 98
Nitrogen fixation, 98
Nondisjunction, 85
Nuclear membrane, 45
Nucleic acids, 29
Nucleolus, 41, 45
Nucleoplasm, definition, 64
Nucleoproteins, 32–33
Nucleoside, definition, 29
Nucleotide, definition, 29
Nucleus, definition, 21

O

Olfactory nerves, 135
Oparin, A. I., 21
Optic chiasma, 134
Optic stalk, 135
Orders, in nomenclature, 14
Organic compounds, 24–26
Organs:
  definition, 48
  presence of, in classification of animals, 145
Osmosis, 100–2
Osseous tissue. See bone
Ovary, 78
Oxidation-reduction mechanism, 107
Oxidation-reduction reaction, 24
Oxides, 23
Oxygen cycle, 97

P

Pachytene, stage of meiosis, 76
Pancreas, 106
Paracelsus, 19
Parasites, 151, 153
Parathyroid glands, 138
Parenchyma, 50
Parthenogenesis, 85
Pasteur, Louis, 20, 63
Pauling, Linus, 70

Pelecypoda, Class, 167
Peptide bond, 28
Peptides, 27
pH, 97
Phenotype, 81
Phoronida, Phylum, 169
Phospholipids, 45
Photosynthesis, 34–35
Phyla, definition, 15
 in nomenclature, 14
Pituitary gland, 116
Plasma cells, 55
Platyhelminthes, Phylum, 157–59
Polarity, in cells, 47
Polypeptides, 27
Population genetics, 92–95
Porifera, Phylum, 154–55
Preleptene, stage of meiosis, 76
Priapulida, Phylum, 165
Prophase:
 in meiosis, 76
 in mitosis, 64
Prosencephalon, 129
Proteins, 26
 formation of, 27
Protistans, 38–40
Protons, 22
Protoplasm, 64
Protozoa, Phylum, 147–53
 classification table, 152–53
Pseudostratified epithelium, 51
Punnett square, 80

R

Recessiveness, in heredity, 79–82
Redi, Francesco, 19
Reduction, definition, 24
Reduction division, 78
Replication, of DNA, 69–70
Reproduction, 111–24
 cellular, 63–78
Reticular fiber, 55
Rh factor, 91
Rhombencephalon, 129
Ribonucleic acid. See RNA
Ribosomes, 46

RNA, 30, 65, 68–73
Rondelet, Guillaume, 17
Rotifera, Class, 160

S

Saccharide, 26
Saccular glands, 53
Scaphopoda, Class, 168
Schleiden, Matthew, 41, 63
Schwann, Theodor, 41, 63
Scientific method, 13
Segregation, law of, in genetics, 84
Servetus, Michael, 16
Sex determination, chromosomal, 84
Sexual reproduction, 111–13
Simple epithelium, 50–51
Single-celled organisms, 37–40
Sipunculoidea, Phylum, 165
Sol state, 32
Spallanzani, Lazzaro, 20
Species, in nomenclature, 13
Sponges, 48
Spontaneous generation, 19–20
Sporulation, 73
Squamous epithelium, 50
Staining, histologic sections, 49–50
Stomach, 105, 139
Strasburger, Eduard, 63
Stratified epithelium, 51–52
Stratum corneum, 52
Stratum germinativum, 52
Stroma, 50
Supportive tissues. See Connective
 tissues
Surfaces, chemistry of, 72
Symmetry, in classification of ani-
 mals, 145
Synapse, 131

T

Taxonomy, 13
Telencephalon, 134
Telophase:
 in meiosis, 78
 in mitosis, 65

Testis, 78
Thymus gland, 138
Thyroid gland, 137
Tissues:
 complexity of, as method of clas-
  sifying animals, 145
 definition, 48
 types of, 50
Tonsils, 137
Transfer RNA, 70
Transitional epithelium, 51
Translocations, 89
Trematoda, Class, 158
Turbellaria, Class, 157
Turner, William, 17

U

Unit membrane, 100

V

Vacuoles, 42
Vagus nerve, 136
Variation, in genetics, 83–84
Venous system, 140
Vertebrata, table of classification of,
 175
Vertebrates, 50
Vesalius, Andreas, 16
Virchow, Rudolf, 63
Viruses, 33, 37
Visceral arch, 137
Vitamins, 102–5

W

Weismann, August, 66, 82

Z

Zoology:
 definition, 13
 history, 13
Zygote, 76
Zygotene, stage of meiosis, 76